Edward Hull

The Physical Geology and Geography of Ireland

Edward Hull

The Physical Geology and Geography of Ireland

ISBN/EAN: 9783337322250

Printed in Europe, USA, Canada, Australia, Japan

Cover: Foto ©berggeist007 / pixelio.de

More available books at www.hansebooks.com

THE

PHYSICAL GEOLOGY & GEOGRAPHY

OF

IRELAND

BY

EDWARD HULL, M.A., F.R.S.

DIRECTOR OF THE GEOLOGICAL SURVEY OF IRELAND, AND PROFESSOR OF GEOLOGY
IN THE ROYAL COLLEGE OF SCIENCE, DUBLIN

WITH TWO COLOURED MAPS AND TWENTY-SIX WOOD ENGRAVINGS

LONDON
EDWARD STANFORD, 55 CHARING CROSS
HODGES, FOSTER, & FIGGIS, DUBLIN
1878

[*All rights reserved*]

INSCRIPTION.

To the Right Hon. the EARL OF ENNISKILLEN, D.C.L., F.R.S.

MY DEAR LORD ENNISKILLEN,

To you who have taken so honoured and leading a position amongst the Geologists of Ireland, and have contributed so largely to the advancement of Palæontology, especially in the department of Ichthyology, I venture to inscribe this little volume, with feelings of the highest esteem and regard.

I remain, my dear Lord,
Your faithful servant,
THE AUTHOR.

5 Raglan Road, Dublin:
October 31, 1877.

INTRODUCTORY.

IRELAND has been designated by a distinguished German naturalist 'The Land of Giant Stags and Giant Causeways,'[1] thus graphically naming the natural peculiarities for which this country is widely known amongst foreigners and students of Natural History. For, although the remains of the *Megaceros* are by no means exclusively restricted to Ireland, but occur at intervals over England and Western Europe, they have been found in such profusion in some parts of this country that the specific name *Hibernicus* [2] has been proposed by Professor Owen as pointing to the region of the 'Giant Stag's' favourite haunts, and where it was free to

[1] Dr. Ferd. Roemer commences a short sketch of his tour throughout Ireland in 1876, thus :—'Ich war in diesem Herbst im Lande des Riesendamms und der Riesenhirsche. Schon lange hatte ich gewünscht, die grüne Insel kennen zu lernen.' Neu. Jahrbuch f. Min. Geol. u. Palæon. (1877.)

[2] 'Palæontology,' 2nd edit., p. 405.

roam unmolested by many of the fierce carnivores which infested the lands on the opposite side of St. George's Channel.

Those, however, who have studied the physical history of this country are well aware that it is full of interest to the student of Nature, and that it offers many special problems for solution. The relative ages and mode of formation of its mountains, the origin of its Central Plain, of its river-valleys, and of its numerous lakes; its volcanic phenomena, the evidences of extensive glaciation at a former period exhibited by its rock-surfaces, its eskers and extensive moorlands—all these will repay careful study, and will form the subject of the following pages.

In order to lay a solid foundation for the more recent phenomena to be treated of, it will be advisable first to present a brief sketch of the solid geology of the country. Fortunately, geological science is becoming so generally a subject of study that it is likely to be ere long the property of the many, rather than, as at present, of the few; and along with

Introductory. ix

the extension of knowledge of its general principles there is being diffused sounder views regarding the mode in which the outward configuration of the earth's surface has been brought about. Thanks to the labours of the modern school of geologists, the worthy disciples of Playfair and Hutton (if not of still older philosophers), it is generally recognised that the features of the landscape are due to the action of water, under its varied forms and modes of working; whether as rain, rivers, torrents, ocean-waves, snow and ice, operating on rocks of varying composition, degrees of hardness, and kinds of structure. We now know that the form of the loftiest mountain equally with that of the slightest eminence, the extent of the widest plain equally with that of the narrowest gorge, the rugged coast-line with its bold headlands and deep indentations, or the featureless shore which descends almost imperceptibly into the sea—all owe their existence to the great sculptor, WATER;—slowly, almost imperceptibly, working on the rocks which have been placed within its reach, where they have been elevated into dry land by the action of those terrestrial forces which, from the earliest times in the world's history, have been converting land into sea, and sea into land. Upon such principles I shall endeavour to place before the reader in a connected form the origin of those features of the landscape which have always made Ireland so attractive to the traveller from other lands, and have endeared her soil to her own people.

CONTENTS.

PART I.
GEOLOGICAL FORMATIONS OF IRELAND.

CHAPTER I.
PALÆOZOIC FORMATIONS.

Table of Formations—Cambrian Rocks—Lower Silurian Rocks —Metamorphosed Silurians of the Districts of West Galway, Donegal, and Derry—Upper Silurian Beds—Old Red Sandstone—Carboniferous Beds—Lower Carboniferous Traps—Yoredale Beds and Millstone Grit—Gannister Beds, Middle Coal-Measures—Irish Coal-fields—Permian Beds . 1

CHAPTER II.
MESOZOIC FORMATIONS.

New Red Sandstone and Marl, Rhætic and Liassic Beds—Cretaceous Beds 50

CHAPTER III.
TERTIARY, OR CAINOZOIC, FORMATIONS.

Miocene Volcanic Formations—Trachyte Porphyry—Augitic Lavas—Volcanic Necks and Basaltic Dykes—Pliocene Clays of Lough Neagh 59

CHAPTER IV.

POST-PLIOCENE, OR DRIFT, DEPOSITS.

Lower Boulder Clay—Middle Gravel (Interglacial) Beds, Upper Boulder Clay 78

CHAPTER V.

POST-GLACIAL DEPOSITS.

Mountain Terraces, Eskers, Local Moraines 96

CHAPTER VI.

RAISED BEACHES AND RIVER TERRACES, SEA-STACKS.

Evidences of a raised Coast similar to that of Scotland—Sea-stacks and Caves—Gravel with worked flints . . . 107

PART II.

PHYSICAL GEOGRAPHY OF IRELAND.

CHAPTER I.

MOUNTAINS.

Leading Physical Features—Central Plain—North-west Highlands, Western Highlands, South-western Highlands, South-eastern Highlands, North-eastern Highlands . . . 117

CHAPTER II.

NORTH-WESTERN AND WESTERN HIGHLANDS OF DONEGAL AND DERRY.

Geological Structure and Age—Upper Silurian Rocks . . 123

CHAPTER III.

SOUTH-EASTERN HIGHLANDS OF DUBLIN, WICKLOW, AND WEXFORD.

Geological Structure and Age 126

CHAPTER IV.

SOUTH-WESTERN HIGHLANDS OF KERRY, CORK, AND WATERFORD.

Geological Structure and Age 130

CHAPTER V.

NORTH-EASTERN HIGHLANDS.

Mountains of Mourne and Carlingford—Geological Structure and probable Age—Table of Mountain Groups of Ireland . 141

CHAPTER VI.

ORIGIN OF THE CENTRAL PLAIN.

Geological Formation—Comparative Carboniferous Sections—Original and existing Carboniferous Areas—Plane of Marine Denudation—Ancient Plane Surfaces—Duration of the Period of Denudation 149

CHAPTER VII.

ORIGIN OF RIVER VALLEYS.

Principles which have determined the existing river-courses . 168

CHAPTER VIII.

RIVER SHANNON.

Its Source—Channel—Original Sloping Plain, or 'Plane of Marine Denudation'—Formation of the Gorge at Killaloe 171

CHAPTER IX.

RIVER BLACKWATER.

Its Course—Professor Jukes's Explanation—Objection to it—
Supposed Cause of its Bend near Lismore . . . 176

CHAPTER X.

OTHER RIVERS IN THE SOUTH-WEST.

Origin and formation of their Channels . . . 197

CHAPTER XI.

OLD DRIED-UP RIVER VALLEYS.

Glen of the Downs—The Scalp and Dingle—Keishcorran—Gap
of Barnesmore—Probable Modes of Formation. . . 182

CHAPTER XII.

ON THE ORIGIN OF LAKES.

1. Lakes of Mechanical Origin. 2. Lakes of Glacial Origin.
3. Lakes of Chemical Solution—1. Lough Neagh—Lough
Allen. 185

CHAPTER XIII.

2. Lakes and Tarns in the Highlands of Galway, Kerry, &c. 192

CHAPTER XIV.

3. Lakes due to Chemical Solution—Lough Ree—Lough
Erne—Lough Corrib and Mask 198

CHAPTER XV.

UNDERGROUND RIVER-CHANNELS.

Cuilcagh Hills—Marble Arch of Florence Court Park—Contrast of the Physical Character of the Carboniferous Limestone of England and Ireland—Lacustrine Denudation of
some Lakes 203

CHAPTER XVI.

BAYS AND INLETS IN LIMESTONE DISTRICTS.

Origin and mode of Formation—Those of Donegal, Sligo, Galway, Kerry, &c. 208

PART III.

THE GLACIATION OF IRELAND.

CHAPTER I.

GLACIAL EVIDENCES.

Rev. M. Close's Observations—General Glaciation and Local Glaciation—Glacial Markings on Rock-Surfaces—Effects of Glaciers in Alpine Districts—Bending over of Strata by Glacial Abrasion—Ice-worn Rocks—Transported Blocks— Glaciated Rocks at Lough Conn, Carrick Mountain, and Benmore, or Fair Head, Co. Antrim 211

CHAPTER II.

PERIOD OF GENERAL GLACIATION.

Lower Boulder Clay, or Till—The Central Snow-field and Axis of Ice-movement in Ireland—Origin and Position of the Central Snow-field—Cause of the Ice-movement out from the Central Axis—Local Centres of Ice-movement during the Period of General Glaciation 224

CHAPTER III.

DETAILS OF MOVEMENT OF THE GREAT ICE-SHEET.

District North of the Central Snow-field—Direction of the Ice-movement—District South of the Central Snow-field—Galway District—Border of the Shannon Valley—Eastern Shores of Ireland—Mourne Mountains—Carlingford Lough a Fiord

—North-eastern District—Relations of the Ice-movement in the North of Ireland to that in the South of Scotland—Ice-movement in the South of Ireland 236

CHAPTER IV.

DEPTH OF THE ICE-SHEET.

Observations of Mr. J. F. Campbell and others—Ice-marks on the Twelve Bins of Connemara—The Reeks—The Devil's Bit Mountain—Wicklow Mountains 260

CHAPTER V.

LOCAL GLACIAL CENTRES OF A LATER PERIOD.

Donegal Highlands—Mountains of W. Galway—The Reeks— The Commeragh Mountains—the Wicklow Mountains—Desirableness of more detailed Observations . . . 263

CHAPTER VI.

A CLOSING CHAPTER.

Extinct Mammalia of Ireland 267

APPENDIX I.

List of Authors quoted 273

APPENDIX II.

List of the characteristic fossils of the Geological Formations of Ireland 276

APPENDIX III.

Geological Maps of Ireland 281

INDEX. 283

PART I.

GEOLOGICAL FORMATIONS OF IRELAND.

CHAPTER I.

PALÆOZOIC FORMATIONS.

THE formations present in Ireland belong chiefly to the oldest and newest periods of geological time as represented in the British Isles. The Cambrian, Silurian, the Old Red Sandstone,[1] and the Carboniferous formations are largely developed, as also the Cretaceous, Miocene, Pliocene, and Post-pliocene beds in the north-eastern portion of the country; but between these two groups there occurs a wide gap in the series of formations—the Permian, Triassic, and Liassic strata being only sparingly represented, and the whole of the Jurassic (or Oolitic) series being without any representatives whatever. This absence

---

[1] The Old Red Sandstone is the freshwater, or lacustrine, representative of the Devonian marine formation. In Ireland the formation on occurs under the former type.

B

of so large a portion of the Mesozoic[1] strata, which are to be found occupying an important position in the geological structure of England, is probably due to *absence of deposition* rather than to *denudation*, by which term we mean the sweeping away of strata which had once been formed. We may, therefore, suppose that at the close of the Carboniferous period the Irish area, with the exception of the north-east (and that only to a small extent), was elevated into dry land, during which time the sea, upon whose bed the Mesozoic strata were deposited, overspread the central portions of England.

In order to bring the representative series of formations on both sides of St. George's Channel more clearly before the mind of the reader, and to indicate their analogies and differences, the following table of strata has been drawn up. This may also be useful for reference in future portions of this work in case the reader's acquaintance with geological terms should happen to be in any degree defective: a contingency, I admit, which is exceedingly improbable:—

[1] 'Or 'Secondary.'

TABLE OF FORMATIONS.

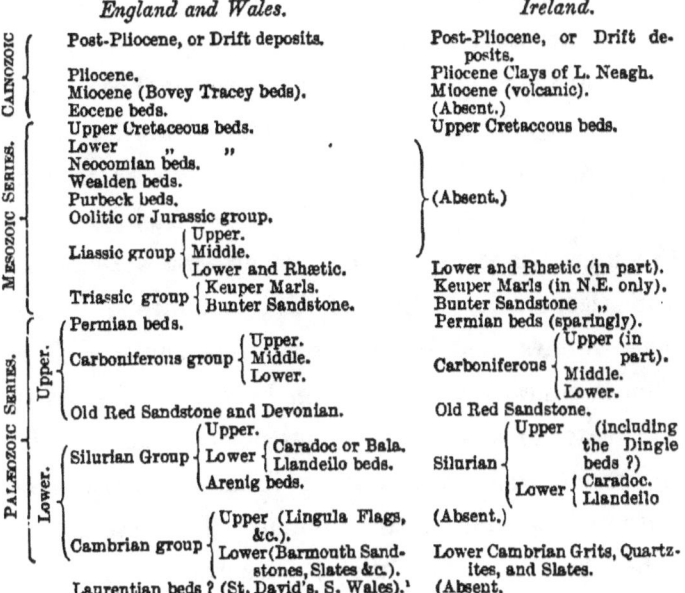

With the above table before him, the reader will not have much difficulty in following me in the attempt to give a concise account of the formations, and their distribution in Ireland; and having thus become acquainted with their petrographical characters, he will more easily understand the action of those causes which have operated in moulding their physical features, and determining their mutual relations over the surface of the country. We shall,

[1] Though the existence of representatives of the 'Fundamental Gneiss,' or so-called Laurentian beds, is strongly advocated by Mr. Hicks and supported by others, their views are not as yet generally adopted.

therefore, commence with the Cambrian Rocks, which are the oldest, and follow the course of geological history downwards; noting those pages which are most fully recorded, as well as those which we can only fill up by having recourse to the geological records of other countries. We shall then proceed to that part of our subject which is really the more important from our present standpoint, and discuss the origin of the mountains, plains, river-valleys, and lakes. All these subjects, it is almost unnecessary to observe, have been discussed, in greater or less detail, by previous authors; a list of whom, with references to their works, will be found at the end of this volume. To these I must refer the reader for fuller information than can be admitted into these pages. Irish geologists, though, perhaps, not so industrious with either the pen or the hammer as the materials which their country has furnished seem to demand, have by no means been wanting in efforts to elucidate the physical structure of their country. The pages of the Transactions and Journal of the Geological Society of London, and of the Geological Society of Dublin (now the Royal Geological Society of Ireland), bear witness to the labours of Berger, Portlock, Bryce, Oldham, Baily, Jukes, Kinahan, Close, Harkness and others; while the explanatory memoirs which accompany the maps of the Geological Survey, written by the officers of the Survey, afford ample details of the geology of special districts. But amongst the corps of 'knights of the hammer' of which Ireland can boast, there is one whom all will allow to be *facile princeps*; who, though amongst the early explorers of the geological treasures of his country, and the first to reduce to order and system its tangled skein of strata, still survives to enjoy his well-merited honours and to 'fight his battles over again'—battles, however, not with men but with the forces of nature. Sir Richard Griffith will always be remembered amongst men of science as the first to construct a geological map of Ireland; doing for this country what William Smith accomplished for England, and Macculloch and Boué for Scotland. Griffith's map,[1] constructed with the help of the late Mr. Kelly and others, but chiefly by his own personal exertions, must be considered as remarkable for the accuracy with which the boundaries of the geological formations are delineated, the broad views it evinces in the grouping of the formations, and the effectiveness of its colouring. It is a work which only a man of uncommon industry, of great physical powers, and of high intellectual

[1] Published by Hodges, Foster, and Co., Dublin (1855). Jukes's 'Geological Map,' on a smaller scale (1867), is taken partly from Griffith's map, and from the maps, when published, of the Government Survey. A new edition has just been issued (1878).

attainments could have accomplished. The author of the first geological map of Ireland has thus established his claim to be known amongst geologists in all time to come as 'the Father of Irish geology.'[1]

Cambrian Rocks.—The beds which have been referred to this formation are only found along the eastern coast, and in the detached localities; namely, Howth Hill, on the north shore of Dublin Bay; Bray Head, and the coast nearly as far as the town of Wicklow; and along the coast of Co. Wexford, from Cahore Point southwards. They consist of green and purple grits, quartzites, and rough slates, which occur generally in a highly disturbed and broken position, and are of great, but unknown, thickness. Fine sections in these beds are laid open along the Dublin and Wicklow Railway, where it has been carried along the face of the steep and lofty cliffs of Bray Head, sometimes spanning chasms descending below the restless waters, at other times cutting deep into the projecting rocks, or traversing them in tunnels. To the north of Dublin Bay, the old rocks are covered over on the land side by beds of Carboniferous limestone, which appear to have been deposited against their flanks, at that point forming a steeply shelving coast, or possibly sunken rocks in the sea of

[1] As William Smith, who constructed the first geological map of England, is generally known as ' the Father of English geology.'

Cambrian Rocks. 7

the Carboniferous period (see fig. 1); but to the southwards, in the districts of Wicklow and Wexford, they are overlaid unconformably by the Lower Silurian beds, which fold round to the eastward and occupy the coast-line southwards from Wicklow.

The organic forms yielded by these beds are few, and of low types of animal life: they consist of

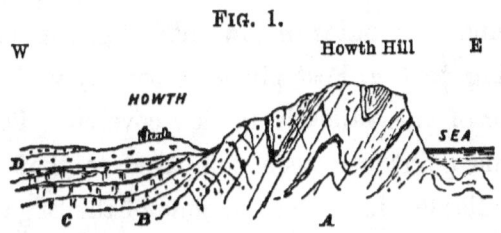

FIG. 1.

A, Cambrian quartzite, &c. B, conglomerate at base of Carboniferous limestone (shore beds). C, Carb. limestone. D, Drift deposits.

the tracks of marine worms, and of a peculiar genus of zoophyte called, after its discoverer,[1] *Oldhamia antiqua* (Forbes).

The surface features of the Cambrian Rocks of Wicklow are generally bold and varied. The beds of quartzite run in rugged and high ridges, or form conical hills, such as the Sugar Loaf Mountains, which are an essential feature in the landscape of that beautiful county. Seen from the north—as for

[1] Dr. Oldham, F.R.S., late Superintendent of the Geological Survey of India.

example, from Killiney Hill—the quartzite ridge appears to trend from east to west, culminating in the stately cone of the Great Sugar Loaf, at an elevation of 1,659 feet, and abruptly terminated along the bold bluffs of Bray Head. From the base of this ridge the wide and richly wooded valley of the River Dargle slopes gently down to the shore of Killiney Bay, which sweeps inwards in a graceful curve, and terminates in low cliffs of gravel and clay belonging to the Post-pliocene period, with which the floor of the valley is thickly covered. The view from this point has been compared to that of the Bay of Naples, in which the cone of the Sugar Loaf represents that of Vesuvius. But the comparison is illusory or fanciful; each possesses features peculiar to itself, and the comparison has probably had its origin in the popular notion that all conical mountains are volcanic.[1]

Lower Silurian Rocks.—In Wales the Lower Cambrian beds, whose representatives in Ireland have just been described, are overlaid by a series of fossiliferous strata now generally denominated *Upper Cambrian*, including the 'Lingula Flags'[2] and

---

[1] Professor Geikie has shown, in his 'Scenery and Geology of Scotland,' how frequently in Scotland formations of quartzite assume the conical form—as in the case of Scheballion; other examples will occur when we come to speak of the western districts.

[2] The Lingula beds, with their peculiar forms of *Trilobites* and a

'Tremadoc slates,' with a peculiar fauna consisting mainly of Brachiopods and Trilobites. These beds, however, have not been discovered in Ireland, and there is every reason to suppose that they are unconformably overlapped, or concealed, along the margin of the Lower Cambrian beds, or have not been deposited over the Irish area. The Lower Silurian beds of the country are therefore the representatives of higher beds in the group known as the 'Llandeilo' and 'Caradoc' (or Bala) beds, which rest discordantly on the Lower Cambrians of the south-eastern coast, and then stretch away northwards, westwards and southwards, forming the general floor upon which the more recent formations have been laid down.

Two Types.—The Lower Silurian beds of Ireland occur under two types, petrologically different, though strictly representative and coeval; and as this is an essential feature in the structure of the country, and must always take an important place in any sketch of its physical history, it will be necessary to dwell upon it for a few moments, and to consider the formation under its two aspects.

If the reader will look at a map of the British

few Brachiopods, were considered by Barrande and Murchison properly to form the base (*zone primordiale*) of the great Silurian series. Professor Ramsay and many other authorities assume the overlying 'Arenig beds' as the base.

Islands (a geological map is preferable), and draw a line from Galway Bay on the south-west to Belfast Lough on the north-east, and then continue it in the same direction through the entrance to the Firth of Clyde, and right across the central valley of Scotland to the Firth of Forth, he will find that (generally speaking) the mountainous regions of both countries lying to the north and west of this line consist of crystalline strata, while the less elevated tracts to the south and east of the line consist of Silurian strata in their ordinary condition, and which are frequently stored with organic remains. In the case of Ireland, the former class of rocks (or the crystalline) form the mountainous districts of West Galway, North Mayo, the Highlands of Donegal and Derry; and in that of Scotland, the central Highlands from the flanks of the Grampians northwards; while the latter class (the ordinary fossiliferous beds) in the south of Ireland rise into a group of isolated mountains such as Slieve Boughta, the Devil's Bit, Galtymore, and Slieve Bloom; and further north into the rugged uplands of Monaghan, Armagh and Down, represented in the south of Scotland by the Southern Uplands. The phenomena of both countries are strictly analogous, and admit of the same explanation. The geological age of the beds south and east of the line I have indicated has been clearly established, both by their fossil contents and geological position: it is recognised that they belong to the great Lower Silurian series of Murchison. But on passing north and west of the 'fateful' line, they undergo a singular transformation. In Scotland these beds are concealed under the Carboniferous beds of the central valley, and in Ireland they also pass under a tract of Carboniferous limestone and other beds newer than themselves, and on emerging in both countries, and rising into the mountainous regions which characterise the northern tracts of the British Isles, they are found to occur in the form of crystalline schists, quartzites, gneiss and granitic rocks, all of which are included by geologists under the general term of 'Metamorphic.'[1]

(1) *Unaltered Silurian Beds.*—The Lower Silurian rocks in their normal condition, to the south of the line where metamorphism sets in, consist in the lower part of dark schists with graptolites, cleaved, and interstratified with lenticular bands of grit, the whole representing the 'Llandeilo Beds' of North Wales; and in the upper, of grey and greenish slates,

---

[1] The metamorphism of strata is doubtless due to the action of intense heat, in the presence of aqueous vapour or superheated steam, accompanied by pressure. It has taken place at considerable depths below the surface; and thus shales and slates have been converted into mica-schist, hornblende schist, &c.; grits and shales into schist, gneiss and granite; sandstone into quartz-schist and quartzite.

grits and fine conglomerates, in some places calcareous and fossiliferous, and in a few localities containing beds of limestone, as on the coast near Donabate, and at the Chair of Kildare. These are the oldest fossiliferous limestones in Ireland, and may be considered as the representatives of the celebrated 'Bala Limestone' of North Wales.

Throughout the north-eastern district these strata are thrown into numerous flexures, and often reversed foldings, the axes of which run in ENE. and WSW. directions, corresponding to the trend of these beds in the Southern Uplands of Scotland. In the districts north of Dundalk Bay, and flanking the shores of Carlingford Lough, they have been penetrated by granitic and plutonic rocks, forming the Mourne and Carlingford mountains, while along a line of country extending from Slieve Gullion to Slieve Croob they have been converted into granite by a process of intense metamorphism.

It is also through these rocks that the granite of the Dublin and Wicklow mountains has been developed, apparently by the metamorphic process, accompanied by eruptive outbursts, giving origin to dykes and protrusions of the granite into the surrounding schists. All along the margin of the granite, the Silurian rocks are highly altered, being converted into micaceous schists, with chiastolite and other minerals which are frequently found in similar positions. To the question of the geological age and origin of the mountain ranges here described we shall return in a future page.

Another point of analogy between the Lower Silurian beds of Ireland and those of Wales is the occurrence of extensive sheets of felspathic trap, ashes, and agglomerates, which were ejected from submarine volcanoes at intervals during the deposition of the sedimentary strata. These beds of volcanic materials are to be found in the districts of Waterford, Wexford, Wicklow, and Louth. They consist of felstones, porphyries and elvanites,[1] which are interbedded with the grits and slates of the formation; and it is probable that the vents from which they were poured out were in active eruption at the time when the Old Silurian volcanoes of North Wales were evolving those great sheets of trap which rise into the grand ridges of Cader Idris, Aran Mowddwy, and the flanks of Snowdon. But while the original materials out of which the physical features of North Wales have been carved, have been somewhat similar in both countries, the scenery of the unaltered Silurian districts of Ireland is seldom so striking and bold as in the former country. Except in the case of the isolated mass of

[1] Porphyries rich in free silica or quartz.

Galtymore, which rises grandly from the central plain to an elevation of 3,015 feet, the Silurian tracts now being described consist of rugged and undulating ground, of no great elevation, and destitute of such striking physical features as those of North Wales. It is in the region of metamorphism, in the Western and Northern Highlands, that the Silurian beds show themselves under their nobler aspect;—but an aspect representing rather that of the scenery of the Scottish Highlands than of North Wales. To this type of structure and outline we shall presently turn our attention.

(2) *Metamorphosed Silurian Beds.*—We now proceed to consider these rocks as they occur under their metamorphosed condition in the west and north-west of Ireland. They occupy four distinct tracts, separated from each other by bays and arms of the sea, which seem to have originally been depressions filled in with Old Red Sandstone and Carboniferous rocks, but which have since been partially removed by denudation and their place occupied by the waters of the Atlantic Ocean. These tracts consist of:—first, the district of West Galway and part of Mayo lying between Galway Bay and Clew Bay, bounded on the east by the great chain of lakes, Loughs Carra, Mask and Corrib; on the west by the Atlantic, which often penetrates deeply into the land

Metamorphic Silurians.

in long arms or fiords, such as that of Killary Harbour. Second, the district of North-west Mayo, lying to the north of Clew Bay and bounded inland by the Carboniferous district of Erris, and by the long tongue of Old Red Sandstone which stretches along the coast of Clew Bay to the base of Nephin. The third is the district of the Ox Mountains, which stretches in a long ridge flanked by Carboniferous rocks, in a NE. direction from the shores of Lough Cong to those of Lough Gill; while the fourth and most extensive is the region lying between Donegal Bay and Lough Foyle, presenting its northern and western face to the Atlantic, and bounded inland by tracts of Carboniferous rocks or of Old Red Sandstone, which sweep round in a broken semicircle from the shores of Donegal Bay to those of Lough Foyle. This district embraces the greater portion of the uplands of Londonderry.

West Galway District.—This district includes that well-known group of mountains known as the ' Twelve Bins of Connemara,' which rising from its centre in bold relief, and culminating in Benbaun, 2,395 feet, are remarkable for their individuality, and conical or dome-shaped forms. They are composed of beds of quartzite, rising in great arches, or folds, from beneath the schistose rocks, which stretch away to the southward, and are traversed by many faults, or fractures, hewing (as it were) the masses of quartzite into rude blocks, which Nature, the great sculptor, has moulded into the forms they now present. Some of these fractures traverse the group from north to south, and coincide with the wildly beautiful valley of Lough Inagh, flanked on one side by Bennabeola, and on the other by Benbaun and the isolated mass of Lissoughter. The sides of these hills are sometimes perfectly destitute of vegetation, their dry gritty substance offering but little footing for even the saxifrage or the fern; they have also undergone considerable polishing from former glacial action; so that it will be easily understood how, seen from certain directions and under favourable sunlight, the mountain sides glisten like glass, or rather with the rich yellowish hue of burnished gold.

The quartzite beds of the Twelve Bins dip to the southward beneath a series of hornblendic and micaceous schists with bands of crystalline limestone and serpentine, which furnish the variegated green ornamental stone known as 'Connemara marble.'[1] Further south and extending along the shores of Galway Bay there is a granitic region, consisting chiefly of the porphyritic reddish granite of me-

---

[1] This is really an ophi-calcite, or mixture of serpentine and calcite in irregular bands of varying hues. The principal quarry is at Recess near Glendalough.

Quartzite Mountains.

tamorphic origin, amongst which masses of eruptive granite and other igneous rocks have been intruded.

To the northwards of the Twelve Bins the quartzites and schists pass unconformably below beds of Upper Silurian age, which extend along both banks of Killary Harbour, forming the heights of Muilrea and Bengorm; they again emerge to the north of this tract, stretching to the shores of Clew Bay, from which rises the quartzite ridge of Croagh Patrick, 2,510 feet at its summit above the waters of the bay (fig. 2),

FIG. 2.

Croagh Patrick, a quartzite mountain 2,510 feet high, in the west of Ireland, as seen from the south-west (after a sketch by Mr. S. B. Wilkinson).

and which when seen from the east or west, in the direction of its axis, has the appearance of a perfect cone, sharply pointed and perfectly symmetrical. From the summit of this cone a wonderfully diversified panorama may be observed. To the east, the eye wanders over the low-lying central plain of

Ireland, its northern margin well defined by uprising hills. Along the north the waters of Clew Bay wash the base of the ridge, and looking across, the eye rests on the rugged moorlands of the Ox Mountains giving place to the broken ridges of quartzite which range through Erris, amongst which rises, in solitary isolation, the quartzite dome of Nephin. To the left are the ridges of Curraun Achil on the mainland, the pointed crest of Slieve More in Achil Island, and the bold headlands of Achil Head and Croaghaun. To the west the Atlantic stretches away to the horizon, studded by several islands, bordering the coast; Clare Island, at the entrance to Clew Bay, being the most conspicuous. Turning towards the south, a great expanse of broken moorland stretching from the base of the ridge is bounded by the terraced slopes of the Upper Silurian mountains, which extend from Muilrea (2,688 feet) on the right, by Bengorm to the tableland of 'Slieve Partry'; on the left, intersected by the valley of Doo Lough, the cleft of Delphi and the Erriff valley. Behind this ridge, the eye may catch the bright glistening summits of some of the quartzite mountains of Connemara, which contrast strongly with the darker masses of the Muilrea Mountains, composed as they are of massive beds of grit, slate, and trap, which stretch across the nearer field of view.

The district of the Ox Mountains does not require

much description. It consists of alternating beds of granite or gneiss schist (sometimes calcareous), and quartzite. The granite is often largely crystalline and porphyritic. In this district we have examples of the results of metamorphic action in the production of varieties of crystalline rocks depending on the characters of the original strata. Thus beds of foliated granite, schist, quartzite and crystalline limestone are to be found succeeding each other in rotation, their characters under their metamorphosed condition depending in all probability upon their original composition, whether as sandstone, shale, grit, or limestone.

Donegal and Derry District.—The rocks of this district may be considered as a prolongation of those of the Highlands of Scotland. They are of Lower Silurian age, but without the basis of Cambrian sandstone and conglomerate, on which the representative beds rest discordantly in Sutherlandshire. Nowhere in the Donegal, Mayo, or Galway districts does the base of this great series of schistose rocks reach the surface; but in Scotland it is otherwise. The geological age of the rocks of the central Highlands has now been placed beyond question by the researches of the late Sir R. Murchison, and the evidence adduced by him has been so fully confirmed by the observations of Professors Ramsay, Geikie, and other observers that it is unnecessary for me here to do more than refer to their writings.[1]

The lowest beds of the district we are now considering occupy the mountainous tract which borders the Atlantic, and which breaks off in a series of grand headlands along the northern coast. The ridges which form this district range in a north-easterly direction parallel to the remarkable valley of the Gweebarra and Owenbarra, which forms a straight depression in the granitic region, somewhat similar in direction, but on a smaller scale, to that of the Caledonian Canal, which traverses the Scottish Highlands. The rocks of this district consist of granite (partly intrusive, but chiefly metamorphic), gneiss, and chloritic or hornblendic schists, with crystalline limestones, and quartzites; the latter being a prominent feature in the physical structure of the district, and culminating in the isolated summit of Errigal at an elevation of 2,466 feet. These beds cross Lough Swilly, and reappear in the promontory of Inishowen, dipping generally towards the S.E. in the direction of Lough Foyle, but subjected, according to Professor Harkness, to several reversed flexures which locally alter the dip of the beds.[2]

There can be no doubt that these quartzites and

---

[1] Murchison's 'Siluria,' 4th. edit. p. 15; also 'Scenery and Geology of Scotland,' by A. Geikie.

[2] Quart. Journ. Geol. Soc., vol. xvii.

limestones represent the beds of similar character, which in Ross and Sutherlandshire rest directly upon the Cambrian conglomerates of Suilven and Queenaig. Like them, also, they pass below a great series of hornblendic and other schists, with occasional bands of crystalline limestone, which form the interior of Donegal and Derry, and are traversed by the River Foyle throughout the greater part of its course; they are ultimately covered by the newer formations along the south and east. These higher beds rise into high tracts of moorland in Derry, called the ' Sperrin Mountains,' which attain an elevation of 2,240 feet; but owing to their comparative uniformity of composition, and the absence of such solid masses of quartzite or granite as those which characterise the older beds of North-west Donegal, they are destitute of the strongly marked physical features which make the N. W. Highlands of Ireland so attractive to the student of art, as well as to the physical geologist.

Upper Silurian Beds.—These strata are but sparingly represented in Ireland, as there is reason to believe (from evidence which I shall presently adduce) that at the close of the Lower Silurian period, the Irish area was generally elevated into dry land and subjected to extensive denudation; and it was only upon the re-submergence of the land, and where deep hollows had been worn out in the older rocks, thus admitting of the deposition of sediment, that strata representing the Upper Silurian period were deposited.

In such a hollow were the Upper Silurian beds which lie along the banks of Killary Harbour and the Erriff valley in the Galway and West Mayo districts laid down; and their presence there is highly interesting, as throwing light on one of the most important physical problems of British geology, namely, the epoch of the metamorphism of the Lower Silurian rocks of the west and north-west of Ireland and of the Highlands of Scotland.[1] For, as we have seen, the rocks of all these districts are representative of each other, and have been subjected to the same process of transformation, or metamorphism. Now, as regards the Highlands of Scotland, the metamorphic rocks of the Grampians are overlaid directly by the Old Red Sandstone all across from the Clyde to the shores of the North Sea, and the only inference we can draw from the relations of these beds is, that the metamorphism took place prior to the formation of the Old Red Sandstone itself. But in the district of West Galway, we can show that the epoch of metamorphism must be pushed still further back; and we are able to assert with

[1] This subject has been ably treated by Professor Harkness, in his paper on ' The Age of the Rocks of West Galway,' &c., Quart. Journ. Geol. Soc., vol. xxii.

precision that it is to be placed at that unrecorded epoch which elapsed between the close of the Lower, and the commencement of the Upper, Silurian period.[1]

The basement beds of the Upper Silurian group of this district consist of masses of conglomerate and shingle, with grits and shales, formed from the waste of the metamorphic rocks against which they rest (fig. 3); these latter had thus been transformed, con-

FIG. 3.

Section N.E. of Kylemore—Connemara.

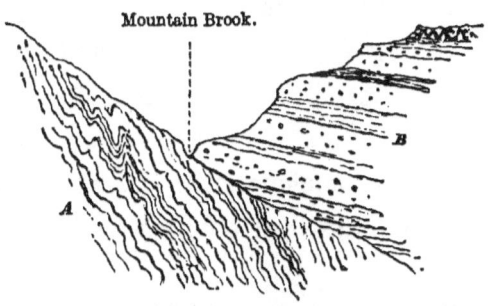

A, Metamorphic schists.   B, Upper Silurian grits, shales and conglomerates resting unconformably on and against A.

solidated, and ultimately upraised as a coast-line on each side of the deep channel in which the shingle now forming the Upper Silurian beds was spread out by the waves and currents of the sea.[2]

Above the basement conglomerates there occurs

[1] Possibly represented, as Professor Harkness has suggested, by the Lower Llandovery stage of North Wales.
[2] These basement conglomerates are full of fossils of Upper Llandovery types, so that their age is beyond question.

a great series of greenish grits and conglomerates, with reddish-purple and greenish shales interstratified, representing probably the Wenlock and Ludlow beds of Shropshire. These stretch from the shores of Lough Mask to those of the Atlantic; and in Muilrea, which stands like a great watchtower guarding the entrance to Killary Harbour, they rise into the highest elevations in the west of Ireland. These beds give evidence of contemporaneous volcanic action; for several sheets of felspathic lava, with accompanying beds of ash, are interposed at intervals amongst the grits. Viewed from the southern shore of Killary Harbour, these beds of lava may be seen cropping out in bold relief along the southern flank of the mountain under which they dip, reappearing in diminished numbers on the northern flank.[1]

It will thus be seen that the Upper Silurian beds have not been subjected to the metamorphic action which has so intensely affected the beds upon which, and amongst which, they have been deposited; they are fossiliferous, and are in their unaltered condition of shales, sandstones, and conglomerates. Moreover, as the rounded pebbles which are found abundantly in them consist of quartzite, mica-schist,

---

[1] See Map of the Geological Survey, sheets 83 and 84, with explanatory memoir.

hornblende-schist, &c., derived from the metamorphosed rocks, it is clear that the metamorphic process had been completed before the newer beds were formed. Here we have evidently a great physical 'break in succession,' during which there were disturbances of the strata, elevations of the beds into land surfaces, and the denudation of the more exposed tracts consequent thereon. Hills and valleys were carved out, and ultimately the whole were depressed beneath the waters of the sea, and the hollows were filled with sediments derived from the waste of the older Silurian lands.

The only other locality of sufficient importance to be noticed, where Upper Silurian beds occur, is the extremity of the Dingle promontory. Here along the coast a fine section is opened out in a series of shales, grits, limestones, &c. highly fossiliferous, which probably represent the whole of the Upper Silurian series of England and Wales, from the Llandovery beds upwards to the Ludlow inclusive. Amongst these are also to be found traps and ash-beds of volcanic origin, which were vomited forth from submarine vents during the process of deposition of the fossiliferous strata.[1] These beds again dip beneath a series of grits, shales and conglomerates,

[1] See Explanation to sheets 160 and 170 of the Maps of the Geological Survey.

several thousand feet in thickness, called by the late Professor Jukes 'the Dingle Beds,' as they occupy the greater part of this promontory. They are apparently conformable to the recognised Upper Silurian beds below, and are overlaid in a highly discordant manner by the red sandstones and conglomerates of the Old Red Sandstone formation. Not having hitherto yielded any fossils, they have been placed by the Government surveyors in a kind of neutral territory, and are provisionally unattached to either of the formations with which they are in contact; but it will probably be admitted that their physical relations seem to suggest their association with the Silurian, rather than with the Old Red formation, to which they are strongly unconformable. The discordance in the stratification of the Dingle Beds and Old Red Sandstone is shown in several of the coast sections of the Dingle promontory, and is represented in the Explanatory Memoir of this district from drawings by the late Mr. Du Noyer.

The Old Red Sandstone.—I have already explained that Devonian rocks, which are the marine representatives of the Old Red Sandstone, do not, as far as we know, occur in Ireland; but, under its lacustrine, or fresh water, type of Old Red Sandstone, the formation is extensively represented in the south-west, and less so in the centre and north. It consists of a great accumulation of sandstones, conglomerates, shales, and slates, which have been deposited below the waters of a large lake (or inland sea) as shown by Mr. Godwin-Austin; or more probably, in two lakes separated from each other by a ridge of Silurian land which occupied the central portions of Ireland; this dividing ridge includes the district of Connemara, and in the opposite direction probably stretched across the Irish Sea to take in North Wales and Central England. The existence of such a ridge will account for the absence of beds of this formation between the base of the Carboniferous series and the Silurian rocks in the counties of Wicklow, Westmeath, Monaghan and Armagh. To the north of this tract the formation was deposited in a basin which stretched into the centre of Scotland; and to the south in a basin the western and southern shores of which lie below the Atlantic waters. The thinning out of these beds in Co. Wexford renders it improbable that the basin was connected with that of the south-west of England and Wales.

The Old Red Sandstone of the south-west of Ireland consists of two members: an upper, which graduates upwards into the Lower Carboniferous Slate; and the lower, consisting of a vast thickness of green, purple, and reddish grits and slates, which rest unconformably upon the Silurian and Dingle beds, as already stated. The formation occupies the greater part of the counties of Cork and Kerry, forming the grand series of ridges which thrust themselves far out into the Atlantic and rise into the highest elevations in Ireland. Their lacustrine origin, notwithstanding their great thickness, is inferred, not only from the absence of marine shells, but from the presence of large bivalves belonging undoubtedly to the fresh-water genus *Anodonta*, discovered in Co. Cork by the late Professor Jukes, and named, after him, *Anodonta Jukesii* (Forbes[1]).

The Upper division consists of finely laminated grey or yellowish flagstones and tiles, rippled, and their surfaces often covered with large worm tracks. They are of no great thickness, but are interesting as containing magnificent fronds of ancient ferns, (*Adiantites Hibernicus*), spread out on the stone as it were with artistic skill, and impressed on its surface in great perfection; other plant-forms, bearing affinity to the future Carboniferous flora, accompany these.

The Old Red Sandstone, after passing below the Carboniferous beds at Killarney and Kanturk, reappears in several detached bosses, which rise from the central plain in dome-shaped masses with cores of

---

[1] Specimens of both the fossil and recent shells, arranged by Mr. W. H. Baily, will be found in the Museum of the Royal College of Science, Dublin. See, Report by Mr. Baily on the Fossils from Kiltorcan Hill, Proc. Roy. Irish Acad., 2nd ser. vol. ii.

Silurian rocks. Between these cores and the margin of Carboniferous beds, the Old Red Sandstone forms an irregular fringe, resting unconformably on the older, and dipping beneath the newer, strata.

The beds of this formation in the northern area have very much the characters of their representatives south of the Grampians in Central Scotland. They occupy an extensive tract of hilly ground from

FIG. 4.

Section near Irvingstown.

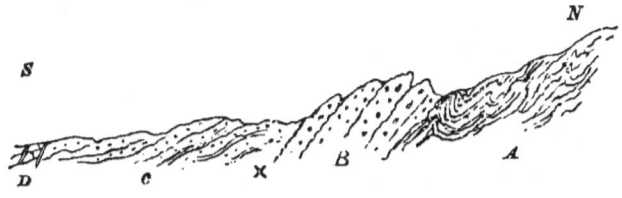

A, Schists.   B, Old Red Conglomerate.   C, Purple sandstone, &c.
×, space unseen.   D, Basaltic Dyke.

the banks of Lough Erne eastwards into Tyrone. Their base, well shown near Irvingstown, consists of a massive conglomerate of large, rounded pebbles of quartzite, felstone, porphyry, and other rocks resting on metamorphic schists, and very similar in appearance to that which occurs along the shores of L. Lomond, L. Long, and the banks of the Clyde. This conglomerate here is succeeded, with some appearance of unconformity, by dark purple, or chocolate-coloured shales, flagstones, and pebbly sandstones, of great but unknown thickness, which have a general SE. dip, though at smaller angles than those of the basement conglomerates. The relations of these beds are represented in the section above (fig. 4). These conglomerates are also well developed on the east coast of Antrim.

Carboniferous Beds.—These beds occupy about one-half the entire area of Ireland, but are represented chiefly in their lowest members, namely, the Lower Carboniferous slate and grit and the Carboniferous Limestone. That formation, which from its physical position in Central England is called 'the Mountain Limestone,' is in Ireland the formation of the plains; except in the north-western districts of Leitrim, Sligo, and Fermanagh, where it rises into bold escarpments and isolated hills. The upper members of the Carboniferous group, including the Coal-measures, occur in a few detached patches, in the north and south; remnants of a once widely extended formation, which probably covered at least three-fourths of the entire surface of the country, but has since been almost entirely swept away by the agents of denudation. To this subject we shall return, when we come to discuss the origin of the great central plain of Ireland.

The basement beds of the Carboniferous group consist in the south of Ireland of cleaved slates and

Lower Carboniferous Beds. 31

grits ('Coomhola grits'), which swell out to a great thickness in Cork and Kerry, and are full of marine fossils (*Cucullæa, Posidonomya*, &c); they are conformable to the Upper Old Red Sandstone. In the centre and north of the country these basement beds consist of yellowish conglomerates, grits and shales, with earthy limestones, representing the 'Calciferous Sandstone series' of Scotland. A fine section in these beds is laid open along the precipitous cliffs which border the Atlantic, west of the entrance to Killala Bay. The thickness of the series must here be about 1,500 feet.

Resting on the basement beds above described comes the Carboniferous Limestone, stretching across the central plain from sea to sea, and northwards to the base of the Donegal Highlands, skirting the mountains of Mayo and Galway, then encircling the isolated domes of Silurian and Old Red Sandstone which rise from the central plain towards the south-west, and finally terminating at the base of the Killarney Mountains along the line of a great reversed flexure, which ranges from Dingle Bay eastwards for many miles. Owing to the numerous foldings into which the Carboniferous and Old Red beds have been bent, the limestone is found filling the centres of several narrow troughs in the south of Ireland, lying along approximately east and west lines, and bounded by anticlinal ridges of Old Red Sandstone.

The Limestone formation generally consists of three members—a lower, middle, and upper; the middle (or 'Calp beds') being generally more earthy than the other two, which consist in the main of pure fossiliferous limestone; the whole formation attains the thickness of 2,500 to 3,000 feet.

On the other hand, in the north-western counties of Sligo and Fermanagh, the formation is entitled to its English name of 'Mountain Limestone,' for in no district of the British Isles are there grander escarpments and terraces than in that which lies between Sligo Bay and Lough Erne, overlooking the southern shores of Donegal Bay. Over this tract the Upper Limestone, resting on the softer Calp series, rises into a table-land, broken off along walls and bluffs of rock, jutting out in bold headlands with scarped faces, from 1,500 to upwards of 2,000 feet above the level of the adjoining ocean.

Close examination of hand specimens, and especially thin transparent sections placed under the microscope, will show that this great calcareous formation consists almost entirely of the shells and skeletons of marine animals, such as corals, crinoids, foraminifera, and molluscs. Even the denser masses which to the eye exhibit no appearance of organic structure, when examined under the microscope in the manner above described, are almost certain to show a field full of animal forms, chiefly those of foraminifera and crinoids. The formation, therefore, notwithstanding its great thickness, must be considered as the work of marine animals, which lived in the waters of the clear ocean of the Carboniferous period, generally far removed from land, and uncontaminated by muddy or sandy sediment. Deep sea-soundings in recent times have shown that similar deposits are in course of formation over vast tracts of the ocean bed of the present day.

The Upper Limestone is remarkable for containing numerous beds, and great amorphous masses of a siliceous material resembling flint, known as 'chert.' Sometimes this material completely replaces the original limestone; and we find corals, crinoids, and other forms which were originally constructed of calcareous material, frequently preserved in chert. Thin translucent sections which I have examined under the microscope have shown that the original forms of crinoids, foraminifera, &c. can be more or less distinctly recognised; so that we may infer that the chert is a product of replacement, by silica from solution, of the original calcareous material of which these structures were composed.[1] The general

[1] On this subject see paper 'On the Nature and Origin of Chert-

Fig. 5.

Section through a part of Co. Sligo.

A, Metamorphic Schists, &c. B, Lower Carboniferous Conglomerate. c, Lower Limestone. d, Middle Limestone, capped by e, the Upper Limestone—solid and cherty.

succession of the Carboniferous Limestone series in the N.W. of the country is illustrated by the adjoining section though a portion of Co. Sligo and Mayo. (Fig. 5.)

The Carboniferous Limestone of the south-west of Ireland, like that of Scotland and Derbyshire, is characterised by the occurrence of volcanic rocks. The rich pastoral plain of Counties Limerick and Tipperary, called the Golden Valley, is diversified by a range of scarped hills composed of volcanic rocks, arranged in a somewhat circular form, outside and beyond which are several isolated bosses rising from out of the limestone plain, both to the north and to the south of the circular range. These latter are also composed of felstone porphyry; and, judging from their position, are clearly the old necks now filled with solid trap, through which the ancient lavas and ashes forming the hills were erupted. On examination it is found that there were two successive outbursts of volcanic materials. During the earlier, felspathic lavas along with quantities of ash and *lapilli* were extruded; during the second, which occurred after a considerable interval, and just at the close of the Carboniferous Limestone period, augitic lavas which have consolidated into basalts and melaphyres were poured forth.

beds in the Carboniferous Limestone of Ireland,' by E. Hull and E. T. Hardman. Abstract, Geological Magazine, Sep. 1877.

The following section drawn through Knock Roe, a hill about 672 feet in elevation, will serve to illustrate the succession of volcanic materials of the earlier outburst. (Fig. 6.)

In this section we have no fewer than ten different varieties, or, we might rather say, the products of ten distinct outbursts of volcanic energy, which were spread over the calcareous bed of the sea, enveloping forests of crinoids, or entombing colonies of hapless molluscs or polypes. First, there came an explosion of ashes and *lapilli* forming the bed (1) directly over the Limestone. This was succeeded by a sheet of basaltic lava (2) (which may possibly be an intrusive sheet of later date); then ash again. After this, probably after a considerable lapse of time, there came a fresh outflow of felspathic lava (4), about 200 feet in thickness, followed (probably at intervals) by repeated eruptions of ashes and beds of trap, the uppermost of which, a columnar felstone with crystals of augite, forms the crest of Knock Roe. The pipe or throat, from which the materials were erupted, was one or more of the isolated bosses of felstone-porphyry, perhaps that of the Knocktancashbane Castle, which rise to the north of 'the lower trap band' not far from the city of Limerick. In this direction, indeed, there are numerous beds of volcanic ashes and agglomerate associated with the beds of

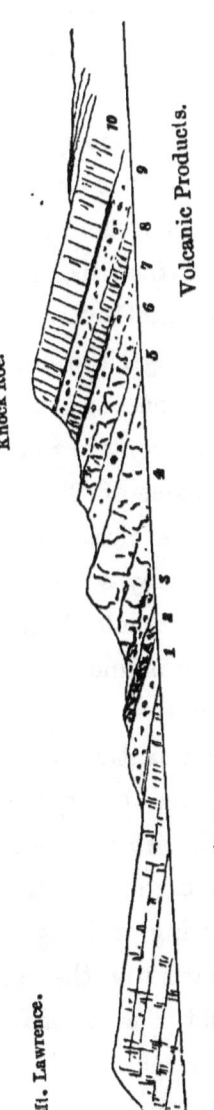

FIG. 6.

Section through Hills of Carboniferous Trap, near Caherconlish, Co. Limerick.

limestone, and by their structure and relations offering objects of great interest to the student of volcanic phenomena.

Yoredale Shales, Flagstones, and Millstone Grit.—The beds which succeed the Carboniferous Limestone consist in the south of Ireland of shales with marine fossils (*Goniatites, Orthoceras, Posidonia,* &c.) surmounted by flagstones ('Carlow Flags'); and in the north of Ireland, amongst the Fermanagh and Sligo Hills, of yellow sandstone and shales, which on the banks of Lough Allen contain rich beds of ironstone, giving rise to the name of Slieve an Ierin (or the Iron Mountain), which rises from the eastern shore of the lake. These may be regarded as the representatives of the Yoredale beds of the north of England, as the name is understood on the Geological Survey, and like them are overlaid by beds of millstone grit, which in the north-western districts rise into high terraced escarpments. Cuilcagh in the Co. Fermanagh, which reaches an elevation of 2,188 feet, is the highest of these. The scenery of the district bordering on Lough Allen, and giving birth to the head waters of the Shannon, resembles in many respects that of parts of Derbyshire, Lancashire, or Yorkshire formed of the same beds. The high table-land of Cuilcagh bears a strong resemblance to 'The Peak' of Derbyshire; and to the north, south, and west, long sweeping terraces of grit, bounded by deep heather-clad ravines, or broad valleys, stretch away towards Donegal Bay, and overlook the wide and richly wooded valley of Lough Erne.

'*Gannister Beds,*' *or Lower Coal-measures.*—This group of strata, which the labours of Phillips in Yorkshire, and of Binney in Lancashire, have so fully elucidated, is characterised by a considerable marine fauna, and also by thin beds of coal. It lies between the Millstone Grit and Middle Coal-measures in the north of England and Wales, and is also well represented in Ireland wherever Coal-measures occur. In the Castlecomer district of Kilkenny and Carlow, it forms a band of strata consisting of hard grits, flagstones and beds of shale, about 600 feet in thickness, with two or three thin seams of coal, and several beds of fossils of marine genera, such as *Phillipsia* (a trilobite), *Orthoceras, Goniatites, Productus, Pallustra, Orthis,* &c. One of the beds of coal has a hard siliceous floor, like the 'gannister' of Yorkshire; and thus in almost every way do these strata correspond to their representatives in England. The Gannister beds also occur near Dungannon on the borders of the Tyrone coal-field, and on the tops of the hills bordering Lough Allen. I have dwelt somewhat longer on these beds than their importance would seem to warrant; but they are of interest from the fact that they form the uppermost member of the essentially marine beds of the Carboniferous group, and their importance in this respect has not hitherto been sufficiently recognised by geologists.[1] To this lower series of beds the little coal-fields of L. Allen, and Kanturk in the County of Cork, belong.

Middle Coal-measures—Coal-fields.—The coal-fields of Ireland belonging to this division are of small extent and productiveness. Being the uppermost beds of the Carboniferous group, they have naturally suffered most from the denudation which has been co-extensive with the whole of the Carboniferous area, and has been so destructive of the upper and most valuable portion, regarded in an economic point of view. They are confined to two districts, that of Coal Island in the north, and that of Killenaule and Castlecomer in the south. As it is not necessary in this place to enter into a detailed account of the structure or resources of the Irish coal-fields, I shall content myself with a brief account of their characters as bearing upon the physical geology of the country, referring the reader for information on the economic aspect of the subject to the works of several authors which are easily accessible.[2]

[1] For further details see Author's paper 'On the Upper Limit of the essentially Marine Beds of the Carboniferous System of the British Isles,' &c. Quart. Journ. Geol. Soc., vol. xxxiv (1877).

[2] Griffith's 'Report on the Tyrone Coal-field,' communicated to

The Tyrone coal-field is of triangular form, being bounded on the north by a large upcast fault, while it terminates towards the south in the direction of Dungannon, and passes beneath the Triassic beds along the east. The 'Lower measures' with the Drumglas coal occupy a considerable area; but at the R. Torrent they are overlaid by the 'Middle measures' with about twelve seams of workable coal. All these pass under the New Red Sandstone towards Lough Neagh, but have not been followed in that direction to any great extent. The coal-field is much broken by faults, and has been worked generally in an unskilful manner, otherwise it might have been rendered much more productive, and beneficial to the district (which is to a large extent a manufacturing one) than has actually been the case; the coal is bituminous.

The coal-field of Killenaule, in County Tipperary, consists of a narrow trough, or synclinal, of measures ranging in a NE. direction, the beds dipping from their outcrop towards the axis at a high angle. It is of small extent, and the coal-seams are thin; nevertheless, they are worked with considerable

the Royal Dublin Society. Kane's 'Industrial Resources of Ireland.' Du Noyer, ' On the Structure of the Lough Allen (Arigna) Coal-fields,' Geological Magazine (1863). Hardman, 'On the Geology of the Tyrone Coal-fields,' Mem. Geol. Survey Ireland (1877), Kinahan and O'Kelly, Explanations to sheets 136, 137, Ibid.

success, by 'The Mining Company of Ireland.' The coal is anthracite, like that of South Wales.

The coal-field of Castlecomer extends over portions of two counties, Kilkenny and Carlow. It occurs in the form of a broad basin, the strata dip-

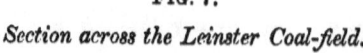

FIG. 7.

Section across the Leinster Coal-field.

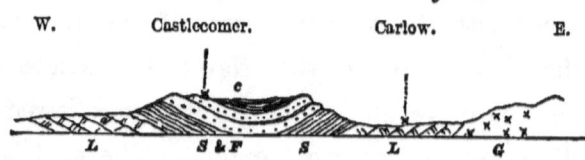

G, Granite. L, Carboniferous Limestone. s and F, Yoredale Shales and Carlow Flags. c, Lower and Middle Coal-Measures in centre of Basin.

ping from the circumference towards the centre. So that the higher and more productive beds of the Middle coal-measures occupy the centre of the basin. The whole district assumes the form of an elevated table-land, rising above the surrounding plain of Carboniferous limestone, over which the beds once extended. The rim of the table-land is generally formed of the 'Gannister Beds,' which rest upon the Carlow Flags, and dip below the upper beds of the centre. The district is traversed by several large faults, which are indicated on the maps of the Geological Survey. The coal, like all the coal of the south of Ireland, is anthracitic. Reptillian remains from the 'Jarrow coal' of peculiar characters have made this coal-field somewhat celebrated amongst palæontologists.[1]

The events which followed the formation of the Carboniferous rocks over the Irish area may now be touched upon. After the Coal-measures had been deposited, the surface must have presented the appearance of a vast lagoon, but little, if anything, elevated above the level of the sea, and (as we may judge from the similarity in the succession of the beds) physically connected with the Carboniferous tracts of England. In the north-west and south-east the older rocks rose above the surface of the lagoon, and formed portions of the unsubmerged lands from which the sedimentary materials composing the Coal-measures were derived. But at the close of the period terrestrial movements, for a long time in abeyance, arising apparently from contraction of the earth's crust in a direction nearly north and south, set in. The forces thus brought into action influenced the whole of the British Isles and the West of Europe, and to them are due the numerous flexures, and lines of geological boundary, which may be observed ranging in approximately east and west directions in these countries. As regards the Irish

---

[1] The Upper Coal-measures, which in Lancashire attain a thickness of 2,000 feet, are not represented in Ireland. If they were ever formed, they have since been completely denuded away.

area, they produced their most powerful and striking effects over the southern and northern districts, while the centre was only partially and locally disturbed and elevated. Over the former districts the beds were forced by lateral pressure to arrange themselves in numerous foldings, very evident, upon an inspection of the Geological Map, by the narrow bands of Carboniferous beds alternating with those of Old Red Sandstone. So powerful, indeed, were the tangential forces that along the base of the Killarney Mountains the beds appear to have been thrust back upon themselves; so that the Carboniferous Limestone of the plain dips beneath the Old Red Sandstone of the mountains of Kerry;—but to this subject we shall return when we come to treat of the mode of formation of the mountains themselves.[1] Meanwhile, as the anticlinal ridges were forced into the air, and as the whole of the central plain was probably somewhat elevated into a land surface, the agents of denudation commenced to wear down the exposed surfaces, and thus began the process of destruction to the coal-formation which was continued through so long a period that (as we have seen) only a few isolated remnants have been left as monuments of the ruin which has overtaken this once widely-spread formation.

The amount of denudation in the north of

[1] See page 130.

Ireland accomplished even before the Permian period properly set in (or at least before any Permian strata were deposited) was very great; for, as I shall now proceed to show, we find beds of this formation resting directly on the Carboniferous Limestone, the Upper Carboniferous beds having been stripped off previous to the deposition of the newer strata. Similar events took place in the north of England, which at this early period of the earth's history was physically connected by continuous land with the north and centre of Ireland.

The formations which remain now to be described are restricted to the north-eastern districts of Ireland, where they are collected within a narrow compass, and generally represented upon a small scale. It is easy to see that the presence of some, or all, of these beds is due to the solid sheets of trap which, during the Miocene period, were poured over this district, affording a protection to the strata that were still, to some extent, preserved. This great sheltering cap has itself been subjected to a large amount of waste. Its original area extended far beyond its present limits; but seeing how the softer Mesozoic strata are grouped closely around its skirts, we may ask ourselves whether, in its absence, evidence of their former presence could have been left sufficient for us to infer that they ever had been there?

Permian Beds.—The representatives of this formation occur only on a very small scale, but they are of much interest as showing that the physical conditions prevalent in Britain during the close of the Palæozoic period were extended over the N. E. of Ireland. The formation is represented by beds belonging both to the lower and upper divisions, the 'Rotheliegende' and 'Zechstein' formations of Germany, the 'Lower Permian' and 'Magnesian Limestone' of England; and these we shall now proceed to describe.

Lower Permian Beds.—The only representatives of this division that we are at present acquainted with are found underlying the city of Armagh. At the time of my visit in 1872 they had not been recognised; but being familiar with the various types of Lower Permian strata in England, I had no difficulty in recognising the nature of the peculiar formation which is laid open to view in the marble quarries a short distance to the west of the city. The upper portion is a boulder deposit, and the lower a limestone breccia; and as the former is overlaid by the Boulder Clay of the Glacial (or Drift) period, both had doubtless been confounded together by previous observers. The breccia had probably been considered as a broken condition of the Carboniferous Limestone. In this quarry, therefore, we have the curious concurrence of two boulder formations, of different and widely separated periods, superimposed one upon the other (see fig. 8). Though somewhat similar in appearance, there is really a difference between them which the practised eye may easily detect; and the divisional line between the two formations may easily be

FIG. 8.

Section of Permian beds at Marble Quarry, Armagh.

B, Boulder clay (Drift). *d*, Permian boulder beds 2 ft. *c*, Red stratified conglomerate, or breccia—resting on *b* compacted sandy breccia of limestone fragments. L, Carboniferous Limestone.

followed along the face of the cliff.[1] The Permian beds are of a deep red colour, rudely stratified, and at the base where they rest on the limestone consist of consolidated breccia of limestone pebbles in a sandy base like the 'Brockram' of the Cumberland district. The blocks of rock in the boulder bed consist of purple grits and felspathic sandstones, sometimes calcareous, which may have come either from the Silurian or Old Red Sandstone districts to the north

---

[1] Professor Ramsay, who subsequently visited the section with me, fully concurred in the identification of these beds.

or west. They attain a size of two feet in diameter; and, though not showing any actual glaciated surfaces, have the appearance of ice-transported boulders.[1] The beds in fact have a strong resemblance to those of Shropshire and Worcestershire, from an examination of which Professor Ramsay was led to infer the prevalence of glacial conditions in parts of the British Isles during the earlier Permian period.

The representatives of the Magnesian Limestone occur in two localities—one near Cookstown, Co. Tyrone, and the other at Cultra, on the southern shores of Belfast Lough. They have been described by Dr. Bryce and Professor King, who has established their correlation with the Upper Permian beds of England by the occurrence in them of shells of the genera *Schizodus, Bakevellia,* &c. It is only at ebb tide that these beds can be observed at Cultra, where they rest directly on Lower Carboniferous shales and limestones.[2]

From these instances it will be apparent that what I have stated above is borne out by the evidence before us, namely :—that before the Permian period there was denudation on a large scale of the recently

---

[1] 'On the Permian Beds of Armagh,' Quart. Journ. Geol. Soc., vol. xxix.

[2] The bands of magnesian limestone were formerly quarried for chemical purposes, so that little is left visible above the sands of the beach.

Permian Beds.

formed Carboniferous strata, embracing a period of long duration during which the rains, and rivers, and perhaps the waves and currents of the sea, were undermining, and carrying away, the materials of which that formation is composed.

CHAPTER II.

MESOZOIC FORMATIONS.

New Red Sandstone and Marl—Triassic Beds.—The beds of this formation occur as a narrow band encircling the basaltic region of Antrim and Derry. Occupying generally low ground and valleys, such as those of the rivers Blackwater and Lagan and the estuary of Belfast Lough, they rest unconformably on all the formations older than themselves, and in the geological classification form the base of the Mesozoic series. The lower portion, or Bunter Sandstone, consists of soft bright red or variegated sandstone, with marly bands; and the upper (or Keuper Marl) consists of bright red and grey laminated marls with irregular bands of gypsum, which are well shown in the railway cuttings between Larne and Carrickfergus. Near this latter place extensive beds of rock salt occur, and are worked by means of vertical shafts. Sections in the Bunter Sandstone are laid open along the north shore of

New Red Sandstone and Lias. 51

Belfast Lough and south of Dungannon, where the beds have yielded remains of fish.

One of the most interesting exhibitions of the New Red Sandstone strata occurs at the quarries of Scrabo Hill near Newtown Ards. The beds are quarried for building purposes, and consist of light yellow and reddish freestone lying probably at the base of the Keuper division. On the surfaces of the beds sun-cracks, rain-pits, and ripple-marks may often be observed, as in the case of the Stourton quarries in Cheshire. They are capped by masses of spheroidal dolerite, and are evidently in proximity to an old volcanic neck (or vent), as the sandstones are penetrated by horizontal sheets and vertical dykes of basalt, intersecting each other in various directions, and seriously injuring the quality of the freestone in immediate contact with them. As the quarries are very extensive, and the excavations expose a large section of the hill-side, there are few places where the phenomena of igneous intrusive rocks can be more advantageously studied than at this spot.[1]

Rhætic and Liassic Beds.—The representatives of

[1] Detailed sections and descriptions of these beds are given in the Explanatory Memoir to accompany Sheet 37 of the maps of the Geological Survey. Another small patch of New Red Marl with gypsum occurs at Carrickmacross in Co. Louth. This is the most southerly place where the formation is represented. The beds have been deposited in an old depression amongst the Palæozoic rocks. Expl. Mem. Sheets 81 and 82, p. 8.

these formations occur together in a few narrow bands at intervals around the base of the Chalk escarpment of Antrim; namely at Larne, Collin Glen, near Belfast, and at Portrush on the northern coast, where they have been indurated by contact with a sheet of intrusive dolerite. At Larne the beds are well laid open on the coast north of the town; the upper belong to the Lower Lias, and consist of light blue, grey, and dark clays with thin earthy limestones, with *Gryphæa incurva, Ostrea Liasica, Ammonites planorbis,* &c. Below these are dark shales with *Avicula contorta* resting on light greenish calcareous sandstone, somewhat pisolitic, belonging to the Rhætic series; below which are the Keuper Marls.

The main portion of the Lias, together with the whole of the Jurassic (or Oolitic) series, is unrepresented, as the Cretaceous beds immediately overlie the Lias. If the Jurassic beds were ever deposited over the N.E. of Ireland, they have been subsequently swept away before the Cretaceous period; there is, therefore, a great unrepresented hiatus in the geological series between the beds just described and the upper Cretaceous strata by which they are succeeded at Larne.[1]

---

[1] For further account of the Liassic beds see papers by Messrs. Tate and Holden in the Quart. Journ. Geol. Soc. Lond., vol. xxiii. and Portlock's 'Geology of Londonderry,' p. 55, &c.

Cretaceous Beds.—The Lower Cretaceous beds being absent, the Upper, consisting of the Upper Greensand and Chalk, are the sole representatives of this great group of strata. These beds, which are soft and easily disintegrated, occur as a very narrow fringe cropping out from beneath the dark basaltic sheets, to which they owe their preservation, and with which they form a remarkable contrast; the whitest of all rocks being in immediate contact with the darkest.

The Upper Greensand is similar in appearance to that of England, as it consists of a greenish calcareous sandstone, soft, very friable, and sometimes pebbly. Its colour is due to small grains of silicate of iron, which under the microscope are seen to be round or oval, and are probably of organic origin.[1] Good sections occur in Collin Glen and at Larne, and their thickness when greatest may be placed at 30 feet; but it is generally less than this.

The Chalk formation which succeeds is precisely similar to the upper portion of its representative in the south and east of England. It consists of compact white limestone with bands and nodules of flint. Amongst these are some of large size, of the shape of sponges, and perforated through the axis, the orifice

---

[1] Similar strata were discovered by the officers of H.M.S. 'Challenger' to be in course of formation over portions of the deep ocean bed; and this occurrence corroborates the views of the late Professor Ehrenberg, that the grains are the casts of foraminiferal shells.

being filled with chalk.[1] Fossils, such as Echini and molluscs, are locally plentiful, but in reality the whole mass is of organic origin. The Antrim chalk is usually sufficiently hard to allow of its being cut into thin transparent slices, which when placed under the microscope show numerous sections of foraminifera, plates and species of Echinoderms and other forms, imbedded in an impalpable calcareous paste. From specimens examined by Prof. T. Rupert Jones and Mr. W. K. Parker the following genera of Foraminifera were determined: *Lituola? Valvulina, Dentalina, Bulimina, Textilaria, Verneuilina, Globigerina, Planorbulina,* and *Pulvinulina.*[2] The sections of flint present somewhat similar forms under the microscope, except that they were less definite.[3]

The thickness of the Chalk formation varies exceedingly, as its upper surface has undergone extensive erosion previous to the overflow of the basaltic covering. Between the basalt and chalk there is generally an irregular bed of flint-gravel, with red ochre, filling the hollows in the surface of

---

[1] These large fossil sponges are locally called 'Paramoudras'—a name not to be found in zoological dictionaries.

[2] Journ. Roy. Geol. Soc. Ireland, vol. iii. (New Ser.) pp. 88-9 (1872).

[3] Zinc has been discovered by Mr. Hardman in the Chalk of Co Tyrone. Ibid., p. 159.

the rock. These flints have undoubtedly been derived from the beds of chalk, which originally were superimposed on the existing beds, and which have probably been carried away in solution by atmospheric waters, while the insoluble flints have

FIG. 9.

Cliff Section of Chalk overlaid by Basalt.   Co. Antrim.

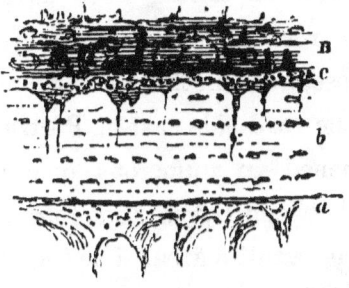

a, Upper Greensand, cropping up from below the Chalk. *b*, White chalk with bands of flint. *c*, Flint-gravel resting on an eroded surface of the chalk. B, Basalt with spheroidal structure.

been left behind. The maximum thickness of the formation is about 350 feet.

In the interior of the chalk-flints there often occurs a white powder, which Mr. Joseph Wright of Belfast has subjected to close scrutiny, resulting in the discovery of a large number of beautifully formed genera and species of Microzoa;—an illustration of the way in which the most apparently worthless objects in nature repay investigation and study.[1]

[1] Trans. Brit. Association (1874), p. 95.

The Antrim coast often displays fine sections in the Chalk capped by dark masses of basalt, or amygdaloid. The beds are generally nearly horizontal, and are often traversed by basaltic dykes, and in some cases by the necks of old volcanoes filled with masses of trap. In other cases the old necks are choked up by bombs which have been blown into the air, and on falling back have filled up the throat. The rock on which stands Dunluce Castle is one of these, and another example may be seen in the cliff by the road-side east of Portrush.[1] Sections are also opened in numerous quarries along the escarpment between Larne and Lurgan, and in the interior of the basaltic area. The Chalk is brought to the surface through the agency of faults at Dunwater quarry and Temple-Patrick.

The occurrence of the Chalk formation so far in advance to the north-west of its position in England, is one of the most curious facts in the physical geology of the British Isles. Were it not for its presence in the north-east of Ireland, it would scarcely have been suspected that it had originally extended into that region. Judging from the abruptly truncated form of its margin, except in the direction of Armagh towards the south-west, it must originally have been spread far beyond its actual

[1] This I now consider somewhat doubtful.

limits. In truth, its composition gives little evidence of proximity to land, for pebbles of older rocks scarcely ever occur, and are only numerous in the Greensand formation at its base. Recollecting that it has undergone two denudations—one before the basaltic sheets were poured over its surface, the other since that time, in conjunction with the basalt itself, we may well suppose that its original margin was far in advance of its present limits. The portion now preserved is in fact the central part of a wide and shallow basin, into which form the beds were tilted after the Miocene period; as they are found to dip generally from the margin inwards. We may, therefore, venture to assume that the formation originally stretched as far as the mountains of Donegal towards the west, and those of Slieve Gullion, Carlingford and Mourne towards the south. In a northerly direction the formation must have wrapped around the flanks of the mountains of Argyllshire, Kintyre, and Arran, and towards the east, those of the south of Scotland and the Lake district of England. If the formation was originally connected with its representative beds in England, which is highly probable, it was so across the area of the Irish Sea; and the depression between the Welsh mountains and those of Cumberland and North Lancashire was probably the line of country through which the connection was made. How far it stretched northwards into the Atlantic Ocean we have no means of knowing, and it imports little to inquire.[1]

From what I have said as regards its composition, it will be evident that the Chalk formation has been built up layer upon layer over the bed of the ocean by the agency of marine animals, for the most part of minute size and low organisation. It contains but little sedimentary matter such as sand or clay, so that it must have been formed far out of reach of estuaries or the mouths of rivers bringing down sediment into the sea. Even the flints are due to organic agency; the silica having probably been secreted by certain forms of animal life from the waters of the ocean,[2] or in some cases replacing the original lime. The representative of this formation at the present day is the foraminiferal, calcareous, ooze of the central Atlantic bed, and of other oceanic regions far remote from land.

[1] Siliceous beds—the representatives of the Cretaceous period—have been observed by Prof. Judd in the island of Mull, in Scotland.

[2] Mr. Wright of Belfast, as above stated, p. 55, has obtained immense numbers of foraminiferæ from the interiors of flints, but these were no doubt originally calcareous.

CHAPTER III.

TERTIARY, OR CAINOZOIC, FORMATIONS.

Miocene Beds; a chapter in volcanic history.—The representatives of the Miocene period in Ireland are not the various forms of sedimentary strata, such as sandstones, shales and limestones found in the centre and south of Europe, but great sheets of basalt and amygdaloid, with beds of volcanic ashes and lacustrine formations of iron-ore and 'bole,' such as are frequently found in similar volcanic regions.[1] On a former occasion, I attempted to present a connected sketch of the succession of events in the volcanic history of this period as it is indicated for us in the north of Ireland,[2] and I shall follow the same course on the present occasion, only with somewhat less detail.

---

[1] The description of the volcanic rocks in the N.E. of India, by Mr. V. Ball, of the Geological Survey of India, might apply, *mutatis mutandis*, to the Antrim district; so great is the similarity of the volcanic products in both countries. Mem. Geol. Survey India, vol. xiii. pt. 2, 1877.

[2] 'Presidential Address to the Geological Section (C.) of the British Association' at Belfast, 1874. Trans. of Sections, Report, p. 60.

FIG. 10.

Showing the N.W. Escarpment of the Basaltic Plateau, with the great Landslips at its base.

1. New Red Marls. 2. Lias. 3. Chalk. 4. Basaltic sheets of Miocene age. 5. Raised Beach.

Tertiary Volcanic Rocks. 61

The area occupied by the volcanic beds, or sheets of lava, extends over nearly the whole of the county Antrim and the adjoining parts of Londonderry and Tyrone, being bounded along the north and east by the sea, and in the other directions by a ridge, or escarpment, marking the commencement of the basalt. The largest lake in Ireland, Lough Neagh, is almost entirely within the volcanic district, and both are drained by the River Bann, which flows northward from Lough Neagh into the Atlantic. The district is generally an elevated plateau, sloping downwards towards the valley of the Lower Bann, and along the north-western and southern borders, overlooking Lough Foyle and Belfast Lough respectively. It is bounded by noble escarpments with precipitous flanks, rising to elevations of 1,000 to 1,560 feet, as represented in the adjoining section taken from one of Gen. Portlock's illustrations. (Fig. 10.)

Along the north coast the scenery is often bold and striking; sometimes, as in the neighbourhood of the Giant's Causeway, the cliffs rise from the sea in a series of terraces of dark columnar basalt, with vertical walls, and separated from each other by bands of reddish bole, or volcanic ash. These great beds or terraces represent successive lava flows, and they differ from one another not only in thickness but in the size and arrangement of the columns. At other times, as at Fair Head directly opposite the Mull of Kintyre, huge columns of basalt descend from the top of the cliff in one or two sheer vertical sweeps for several hundred feet, while at the base of the cliff the shore is strewn with broken columns of trap heaped up in wild confusion;—a Titanic break-water

FIG. 11.

Basalt Cliffs of Fair Head, Co. Antrim, 630 ft. above the sea, seen from the South. Carboniferous beds in the foreground.

which the waves of the sea have reared up against their own advance. The slopes of the escarpment overlooking Lough Foyle are characterised by enormous landslips of trap (see fig. 11), illustrating the progress of atmospheric waste, as those of the coast-line at Fair Head illustrate the destructive effects of wave action. In the former case the great beds of trap surmount softer Cretaceous and Triassic strata (see fig. 11), and as these are undermined by the percolation of water from springs, or by rains, the soft foundations give way, and the heavy superstructure breaks off along lines of jointage, or the faces of the columns; the whole then slips down the hill side and lies a shapeless mass, till it has been still further disintegrated by frost, rain, and streamlet, and carried away particle by particle into the ocean.

The central portions of the basaltic district are often rocky, and near the eastern coast, where indented by valleys running inland from the sea, are diversified by terraced escarpments, but are usually destitute of striking features.[1] In this respect they may be contrasted with the volcanic region of Auvergne in Central France, where the plateau is surmounted by numerous cones and cup-shaped craters, from which the lava streams were originally poured forth, and have spread over the tracts at their base. But in the case of the north-east of Ireland, all the original volcanic cones and craters, though not more ancient than some of those of Central France, have entirely disappeared; and along with them a considerable portion of the sheets of trap on which they were planted. The country has been swept by planing and levelling agents from which Central France has been exempt since the volcanic

[1] In the centre of Co. Antrim the basalt is generally overspread by boulder-clay and gravel.

fires became extinct; hence the contrast in the scenery of the two districts. What these planing agents were we shall enquire further on. I can only here mention their names—the waves of the sea, river-streams, and sheets of ice.

And now let us pass on to consider the nature of the volcanic products of Antrim, and the order in which they were erupted.

As I have shown on the occasion above alluded to,[1] the age of volcanic activity is divisible into three stages, between each of which there was probably a considerable lapse of time, which may be clearly recognised by differences in the volcanic products. The earliest of these is characterised by highly silicated felspathic products, such as trachyte porphyry, pearlstone, and pitchstone; the next, by basic beds of amygdaloid with bands of bole, volcanic ash, &c.; and the newest by more solid sheets of columnar basalt and dolerite, in which the prevalent minerals are augite, Labradorite felspar, olivine, and titano-ferrite (or titaniferous magnetic iron-ore).

(1.) *Earliest stage.*—The lavas of this stage consist of trachyte porphyry, originally described by Dr. Berger in his memoir 'On the Geological Features of the North-east of Ireland' under the name of 'porphyry of Sandy-brae,'[2] and are well laid open

[1] See p. 59.   [2] Geol. Trans. Lond., 1st ser. vol. iii. p. 189.

to view at Tardree and Brown Dod hills. Near the town of Antrim, and in the centre of the volcanic district, the rock has a yellowish felspathic matrix in which are enclosed numerous grains of smoke-quartz and crystals of sanidine. In the little cavities, which are numerous, a very rare form of silica, called 'tridymite,' has recently been discovered by an eminent German mineralogist, Professor A. von Lasaulx of Breslau, who states that it is so abundant in some places that he was tempted to call the rock a 'tridymite trachyte.'[1] This mineral consists of little thin hexagonal plates, generally intersecting in groups of three, and of a yellowish colour; it might easily be overlooked by any but a practised mineralogist. The trachyte rises from beneath the basaltic sheets, being the oldest and lowest of the volcanic rocks, and may possibly belong to the latter part of the Eocene period. At any rate, such is its contrast to the overlying sheets of basalt and amygdaloid of known Miocene age, that I am constrained to infer a considerable lapse of time between their respective eruptions. A similar rock occurs at Temple-Patrick, and west of Hillsborough. These highly silicated and felspathic lavas are of very limited extent as compared with the augitic masses

[1] This discovery was kindly communicated to the Author by Prof. Von. Lasaulx, in a letter written shortly after his return to Breslau in the autumn of 1876.

which succeed them, and they were poured forth over an irregular and much eroded floor of Chalk, or of the still older rocks where this formation (as in the SW. parts of the district) had been entirely removed. In the counties of Armagh and Tyrone the lavas rest on Triassic, Carboniferous, or Silurian rocks indiscriminately.

(2.) *Middle stage.*—The beds of trap representing this stage consist of vesicular amygdaloids and basalt, generally showing a spheroidal structure. Bands of bole (red clays resulting from the decomposition of the trap) are frequent, and the rock is frequently amygdaloidal, containing in its cavities various zeolites and carbonates. They attain in some places a thickness of 600 feet, and are surmounted by bands of bole and volcanic ashes, with great bombs of trap, well seen along the Antrim coast near Dunluce Castle, which mark the close of the second period of eruption. These sheets were erupted probably directly under the air, from vents situated at intervals over the whole area; and this period of volcanic activity was followed by one of repose, marked by the formation of peculiar beds of iron ore and lacustrine plant-bearing strata, which have enabled geologists to determine the geological age of the formation. The plant-remains are seen to best advantage in the stratified ash-beds of

Ballypalidy, and belong to the genera *Sequoia*, *Cupressites*, *Rhamnus*, *Quercus*, *Pinus*, &c., as determined by Mr. W. H. Baily, from specimens discovered by the late Mr. Du Noyer. The whole assemblage indicates a Miocene flora, and is similar to that detected in the basalts of the Island of Mull by the Duke of Argyll in the year 1851.

The bed of pisolitic iron-ore, which is now being extensively worked at intervals over an area ranging from the northern coast near Pleaskin Head[1] to Ballypalidy on the south, is very peculiar, and often rich. It is composed of small grains of red hæmatite, of the size of a pea, or a bean, cemented by red ochreous paste; and is sometimes four feet in thickness. The beds of bole which lie below it are also rich in iron, and exhibit traces of lamination. On the whole, I am disposed to consider these beds as of lacustrine origin, formed in shallow water, and in a depression of the basaltic area due to the sinking of the surface at the close of the second period of volcanic activity. The streamlets which drained the surrounding uplands appear to have carried down the leaves and branches of plants which grew on the margin of the lake. Owing to the decomposition of the basalt their waters also

[1] Where it was originally observed by the Rev. Dr. Hamilton in 1790.

carried down the iron dissolved as carbonate, which on entering the lake was precipitated in the form of oxide of iron. Sections of these deposits may be seen at Port Fad Mine on the north coast, at Broughshane, Shanes' Hill, Glenariff, Ballypalidy, and Island Magee.

(3.) *Latest stage.*—After a period of rest, the volcanic fires again burst forth, giving birth to vast sheets of solid augitic lava, which were poured over the lacustrine deposits just described, and now form successive beds of columnar basalt, which may be studied to great advantage along the cliffs of the northern coast. At Port Fad, and Shanes' Hill near Larne, these upper basalt sheets may be observed rising in rows of massive columns from the upper surface of the iron ore, or of a thin seam of lignite which is sometimes interposed. These upper sheets attain a thickness of about 400 or 500 feet; so that the entire thickness of the lavas may be taken at about 1,100 feet;—a thickness which, owing to denudation, is less than the original, but is much exceeded by the masses of contemporaneous trap in the Island of Mull, estimated by Professor Geikie at 3,000 or 4,000 feet in all.

Volcanic 'Necks' and Basaltic Dykes.—Although the actual craters and cones of eruption have been swept from off the surface of the country by the ruth-

Volcanic Necks. 69

less hand of time, yet the old 'necks' by which the volcanic mouths were connected with the sources of eruption can occasionally be recognised; they sometimes appear as masses of hard trap, columnar or otherwise, projecting in knolls or hills above the upper surface of the sheets through which they pierce, as at Carmoney Hill near Belfast, or that remarkably abrupt mass of trap called Sleamish which

FIG. 12.

Diagrammatic Section through Dunluce Castle, showing the supposed old Volcanic 'Neck.'

B, Amygdaloidal Basalt in sheets, resting on the Chalk formation (C), and penetrated by the old 'Neck' N, filled with agglomerate.

dominates the surrounding country near Ballymena. In other cases, the neck consists of a great pipe choked up by bombs and blocks of trap, more or less consolidated, bombs which have been shot into the air, and have fallen back again. To one of these near Portrush on the northern coast I have already referred, but another not far from this is not so generally known *as such*, namely, the rock in which stands the ruined castle of Dunluce, the ancient fortress of the MacDonnells, its base washed by the ocean. This rock is formed of bombs of all sizes up to 6 feet in diameter, of various kinds of basalt, dolerite, and amygdaloid firmly cemented, and presenting a precipitous face to the sea. The white chalk through which it rises may be seen below the clear green ocean water at low tide, and the coast cliffs, separated from the castle-rock by a broad chasm, are of the same formation. The site must have been a remarkably strong one in the days when cannon were unknown to the warlike tribes of Ulster. The section above (fig. 12) will probably afford a better idea of the relative positions of the rocks at this place than pages of description.[1]

Basalt Dykes.—The whole area of the north-east of Ireland is traversed by basaltic dykes, piercing the different formations in every direction, but more frequently converging towards the region of the

[1] Since the above was written I have had an opportunity of revisiting the coast of Antrim in company with Professor Ramsay and Mr. W. A. Traill; and a further examination of the Rock of Dunluce Castle and of the cliffs adjoining leads us to suspect that we have here, instead of old volcanic necks, simply pipes, formed by filtration out of the chalk, into which the basaltic masses have fallen or slipped down, thus giving rise to their fragmental appearance. Sept. 1877.

basaltic plateau. Some of these are of more ancient date than the Miocene. In the district of Carlingford, where there appears to have been a focus of eruption, most of the dykes are probably of Carboniferous, or Permian, age. To the same periods are referable the large number of trap dykes, which traverse the Silurian rocks along the coast S. of Newcastle at the base of the granite mountain Slieve Donard, but which terminate at the margin of the granite itself, being of earlier date.[1] Similar instances are of frequent occurrence around the granitic district of the Mourne Mountains. But, as Mr. Traill has shown, along with the basaltic dykes of greater antiquity than the granite there are some others which traverse the granite itself and are therefore of more recent origin, and we shall probably not be wrong in referring these latter to the period of the Miocene eruptions of augitic lava.

It is, however, on approaching the base of the basaltic escarpment that dykes of undoubtedly Tertiary age become most abundant. These are found penetrating not only the Triassic and Cretaceous beds, but even the sheets of basalt and amygdaloid themselves. It is rarely, however, that they are found

[1] These are marked in Griffith's Geological Map, and in more detail on the Geological Survey-map, Sheet 61. But so numerous and varied are they that it is only on the 6-inch maps that they can be properly shown.

traversing the uppermost beds of lava except on the northern coast.[1]

It is very instructive to walk along the northern shore of Belfast Lough during ebb tide, and trace the course of the dykes as they protrude in little rocky ridges, running out from the shore into deep water, and cutting across the beds of New Red Sandstone and Marl, which they sometimes convert into a sort of Lydian stone. Some are of large size; such as that on which the solid walls of Carrickfergus Castle have been reared, vying in solidity with the basaltic foundation itself. Similar remarkable dykes, together with intrusive sheets, penetrate the Carboniferous rocks at Ballycastle Bay, where their relations to the strata can be advantageously studied;[2] and if we cross the narrow channel which here separates our island from Scotland, we shall find their representatives along the coast of Kintyre, the islands of Arran and Bute, or the district of the Glasgow coal-field.

Pliocene Clays of Lough Neagh.—Lough Neagh is not only the largest fresh-water lake in the British

---

[1] See account of the Basaltic sheets and dykes of Scrabo Hill, near Newtown Ards, p. 47.

[2] Several narrow dykes traverse the upper as well as the lower sheets of lava east of the Giant's Causeway, leading to the inference that they represent the latest efforts of volcanic activity in the N.E. of Ireland.

Islands, but it is the oldest still surviving. Many of the existing lakes owe their origin to glacial agencies, or to solution of the strata by water, and have been formed during, or since, the Glacial (or Post-Pliocene) epoch. But, Lough Neagh is older than the Glacial epoch; and still survives as a lake, though in diminished size, notwithstanding the physical changes to which it has since been subjected. The evidence of this I shall presently endeavour to give; meanwhile let us examine the proofs of the former more extended range of its waters.

All along the southern portion of the lake, including the eastern and western shores as far as Sandy Bay and Arboe Point, there occurs a tract of country not much elevated above the surface of the lake, (perhaps 80 or 90 feet at the most), formed of grey, purple, and blue stiff clay, with thin laminated sandstones and bands of lignite. In the neighbourhood of Dungannon the clays are used in the manufacture of pots and pans (or 'crocks' as they are called in that country). The clays have been pierced in several places to considerable depths;[1] and Mr. Hardman, who refers them very properly to the Pliocene stage, estimates their maximum thickness

---

[1] In one of these, to a depth of 294 feet, including 30 feet of Drift, at the Townland of Annaghmore. Griffith, 'Second Report to the Railway Commissioners,' p. 22.

at not less than 500 feet.[1] These Pliocene clays have a slight dip towards the lake, and they rest on an eroded surface of the Miocene basalt, their junction with which was very clearly observed not long since by Mr. Hardman and myself along the banks of the River Crumlin, a short distant above its entrance into Sandy Bay. The lowest beds were seen to consist of a conglomerate formed of pebbles of the

FIG. 13.

Section across the Eastern Shore of L. Neagh at Sandy Bay.

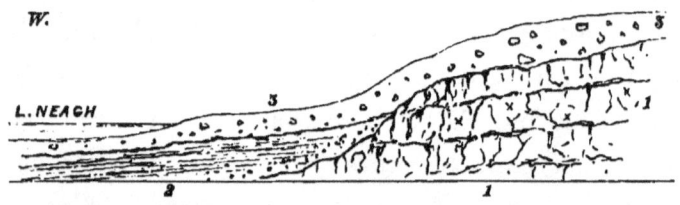

1. Miocene Basalt-sheets. 2. Pliocene clays resting against their flanks. 3. Boulder-clay overlapping both.

basalt, evidently an old shore gravel beach; and then over this were formed the bluish clays, stiff and plastic, in which my companion detected bivalve shells resembling a species of *Unio*, or *Mytilus*,[2] the

---

[1] 'On the Age and Mode of Formation of Lough Neagh,' Journ. Roy. Geol. Soc. Ireland, vol. iv. p. 174–5; Expl. Mem. Sheet 35 of the Geol. Survey Maps, p. 77.

[2] So considered by Mr. H. Woodward, F.R.S., but Mr. Baily thinks they are *Uniones*; if the former be correct, it would show that the sea had access at an early stage of the clay-formation.

first that had been discovered in these beds, or indeed in any British Pliocene strata.

While the clays rest upon an eroded surface of the Basalt, as has been ascertained both by observation and boring experiments, thus proving their age to be more recent than the Miocene stage, they are themselves often covered over by lower Boulder Clay and deposits referable to the Post-Pliocene or Glacial epoch. This is illustrated by the accompanying section (fig. 13), drawn at the eastern shore of Sandy Bay, where the relations of the different formations, namely the Boulder Clay, the Lough Neagh Clays, and the Basalt were very clearly determined by Mr. Hardman and myself on the occasion of our visit in 1876.[1] The clays are, therefore, clearly of an age intermediate between the Miocene of the basaltic sheets and the Post-Pliocene of the Boulder Clay;—in other words, they are of Pliocene age. They appear to be of an age intermediate between one of intense igneous action, and one of intense glacial action; but in their composition they indicate no immediate connection with either. On the other hand, their laminated, silty character, the frequent bands of lignite, and the impressions and bark of plants (*Sequoia, Alnus, Quercus, Fagus,* and *Salix*) indicate

---

[1] See Expl. Mem. to sheet 35 of the Geol. Survey Maps by E. T. Hardman, p. 89.

deposition under tranquil waters, in a warm climate, and with occasional alternations of swampy lagoons choked with vegetation.

That they have been deposited under the waters of what we may call 'Old Lough Neagh' to distinguish it from the existing lake, and probably where an ancient river entered this lake, there can be no doubt; and it is also clear from the wider extension of these beds beyond the margin of the existing lake —especially in the direction of Dungannon and Armagh—that the waters of Old Lough Neagh occupied a considerably wider area towards the south than at the present day. The original banks, or margin, of the lake south of the River Crumlin can be clearly traced, by the abrupt and steep ascent which the basalt produces; and against the base of which the beds of clay have been deposited. Towards the south, however, this original margin is not so clearly definable, it having been formed of the softer strata of the New Red Sandstone, which (along with the clays themselves) are generally thickly covered by a deposit of Boulder Clay. Here, then, we have a formation of an age but sparingly represented in the British Islands, and in which a warm and equable climate prevailed, as indicated by the plant-remains. Ere it set in, the volcanic fires had smouldered away. It was an age of calm repose, separating the period of volcanic activity on the one side from that of frost and ice on the other; and during its continuance the ordinary agents of denudation—rain, rivers, and sea waves—carried on their operations without unusual interruption, but with marked effect in modifying the physical features of the north and adjoining districts of Ireland.

I reserve to a future page the discussion of the question how Old Lough Neagh was formed, and now proceed to describe the characters of the deposits which represent the period succeeding that of the Lough Neagh clays, and known amongst geologists as the 'Post-Pliocene,' 'the Glacial' or 'the Drift,' and which are widely represented in this country.

CHAPTER IV.

POST-PLIOCENE, OR DRIFT, DEPOSITS.

Post-Pliocene, or Drift, Deposits.—These are the newest deposits requiring our attention to any great extent, and are the most widely distributed of any existing in Ireland. It is probably not too much to say, that, in one form or another, they cover three-fourths of the entire surface of the country, resting indiscriminately on all the older and more solid formations, and rising from the plains up the flanks of the mountains to elevations of between 2,000 and 3,000 feet. They are the representatives of a period generally characterised by extreme cold, ' The Great Ice Age,' in which the rains and rivers of the present day were represented by snow, sheets of ice, and glaciers; while the relations of land and sea were subjected to considerable modification as compared with those now prevailing.

The deposits of this period may be arranged under three divisions corresponding with those of the north of England, if not of a much larger area;

namely (1), The Lower Boulder Clay, or Till; (2) the Middle Sands and Gravels; and (3) the Upper Boulder Clay, this being the newest member of the series. These divisions have been identified over a large district extending from Kilkenny on the south to the borders of Tyrone and Londonderry on the north, and may be considered to represent three stages or phases of the Glacial period;—the lower and upper being essentially arctic in character, the middle being temperate. I shall now endeavour to give a short account of the deposits representing each of these divisions in ascending order.

(1.) *The Lower Boulder Clay.*—This consists of a very stiff solid clay, of a dark blue, or reddish, colour according to locality, and containing blocks, pebbles, and fragments of various rocks imbedded therein, and in every possible position. These blocks are of all sizes, either angular or rounded, and often having their sides planed down and covered by scars or groovings, evidently due to their having been forcibly rubbed over hard substances, such as pointed rocks or sharp stones. These scars and scorings are generally well preserved on blocks of limestone which have been freshly disengaged from their beds. The clay itself is very seldom laminated, like those of aqueous formation; on the contrary, it is generally entirely structureless, and the stones and boulders may be observed standing in vertical, or inclined, positions; and not, as in the case of those which have been strewn under water, lying on their flattest surfaces. These peculiarities serve to prove that this remarkable deposit must have been

FIG. 14.

Coast Section near Ardglass, Co. Down, showing the Relations of the Drift-beds and Raised Beach.

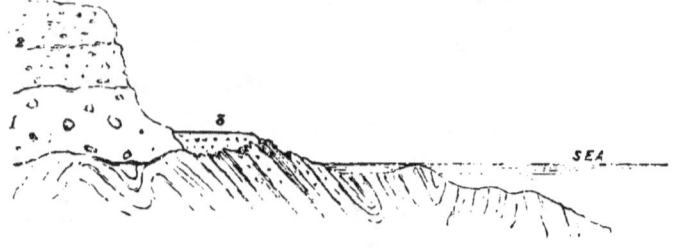

Silurian Rocks.

1. Lower Boulder Clay, resting on a glaciated surface of the Silurian rocks. 2. Middle Sand and Gravel resting on the Lower Boulder Clay. 3. Raised Beach of gravel 15-20 feet above sea.

formed in a manner differing from that in which ordinary beds of gravel or conglomerate have been accumulated; that some agent other than water has been engaged in its construction. Now, the only agent we are acquainted with capable of producing such a deposit is ice, either in the form of a glacier, or of a sheet spread over the country, and moving in certain directions.

This view gains additional confirmation upon observing the surface of the solid rocks from which the Boulder Clay has recently been stripped off. It is frequently found that these surfaces (especially when the rock is hard and compact) are smoothed down into flat, or mammillated forms, and are remarkably polished, or covered by parallel scorings and flutings. In such cases the surface of the rocks is said to be 'glaciated.' Instances of this are of such frequent occurrence that it is scarcely necessary to refer to examples; but in the neighbourhood of Dublin they may be observed along both shores of the narrow neck of land which unites Howth Hill with the mainland, and again to the south of Dublin at Killiney and Ballybrack Hills. They may also be observed along the eastern coast of Co. Down north of Dundrum Bay. In such cases the direction of the grooves and scorings indicates the direction in which the ice sheet has moved.[1]

The tendency of the Boulder Clay to arrange itself in parallel ridges has been noticed by many observers in nearly all parts of the country.[2] These ridges are very well shown on the hill-shaded

[1] It is not my purpose to enter into the proofs that these phenomena are due to the action of ice: the subject will be found sufficiently discussed in various works dealing with physical geology.

[2] A list of these observations has been furnished by the Rev. Maxwell Close. Journ. Roy. Geol. Soc., vol. i. p. 209.

Ordnance Survey maps of the neighbourhood of Dublin, and the Rev. Maxwell Close has clearly indicated their connection with the directions of the ice-scorings of the rocks in their immediate vicinity. This connection will be recognised on referring to the 'Map of the General Glaciation of Ireland,' which accompanies his well-known paper on this subject.[1] As I shall have to return to the discussion of this matter in a future page, I will not further dwell on it at present, except to observe, that the phenomenon here alluded to affords additional evidence that the Boulder Clay is the result of glacial action. These ridges must not be confounded with those of the Esker Gravels, which are due to entirely different causes, and are composed of different materials, namely, water-worn and stratified masses of gravel and shingle, which are of much more recent origin than any of the true glacial deposits, and have been constructed out of these very deposits themselves.

The Lower Boulder Clay is the most extensively distributed of all the Post-Pliocene deposits in Ireland. Being the oldest, it has suffered less from denudation than the more recent Gravels and the Upper Boulder Clay, while it frequently underlies these newer strata, forming the immediate covering of the

[1] Rev. M. Close, Journ. Roy. Geol. Soc., vol. i., plate, p. 224.

solid rocks. It may be observed in this position along the shores and islands of Cork Harbour on the south, as well as along the coast of Antrim and Donegal on the north; and along the borders of the Irish Sea, skirting the coast of Wexford and Wicklow on the east, as along the shores of Galway Bay on the west. Its extension into these districts shows the prevalence of glacial conditions over the whole country, as also indicated by the scorings on the rocks; and thus we have side by side, or in actual contact, two witnesses to the wide extension of glacial conditions at a former epoch in the physical history of Ireland.

As might be inferred from what I have said of its mode of formation, the Boulder Clay occurs in greatest mass in the lower grounds and deeper valleys of the country. The plains and valleys having been in the main formed and hollowed out before the Glacial period, the Boulder Clay naturally filled up the hollows and protected portions of the surface, often choking up the deep gorges, and entirely obscuring the underlying rocks. Many of these valleys have been re-excavated; and the existing streams have worn down their channels to the original solid floors. From the plains and valleys, the Boulder Clay rises on the flanks of the mountains and hills to elevations of upwards of 1,500 feet in many cases, when it either becomes thin and sparingly distributed, or entirely disappears. It is curious also to observe, that on some comparatively low grounds, such as that at Inniskeen in Co. Louth, the old rocks are entirely free from any covering of Glacial drift over a considerable area, while it occurs in great thickness a few miles further east, in the neighbourhood of Castleblaney. Such instances show either the great irregularity in the original distribution of the drift, or the extent to which it has been locally denuded since its formation.

In mountainous districts, such as those of Galway, Mayo, and Mourne, the Boulder Clay assumes the appearance of local moraine matter, made up of angular blocks mainly of local rocks confusedly heaped together with earth or shingle; and it is often difficult, except where fragments of foreign rocks are enclosed, to distinguish such masses from those originating in local glaciers.

(2.) *The Middle Sands and Gravels. Inter-glacial Beds.*—This division is very largely distributed over the central plain of Ireland, constituting what is generally known as 'the limestone gravel,' because largely made up of pebbles of Carboniferous Limestone. It also rises high upon the mountain slopes of Wicklow and Dublin, and is distributed along the eastern coast of Dublin, Wicklow, and Wexford, where it contains numerous species of marine shells.[1]

This formation is very different in character, and consequent mode of accumulation, from the Boulder Clay which it succeeds, and upon which it may be frequently observed to rest. As its name imports, it consists of stratified beds of sand or gravel of water-worn pebbles, sometimes of large size; and, as it contains marine shells in various places, may be regarded as a formation of marine origin, which has been strewn over the bed of a comparatively shallow sea. In these waters local currents appear to have been prevalent, as they have left their traces in the numerous instances we find of 'current-bedding,' or 'oblique lamination.' The fine sections in these deposits along the coast at Ballybrack, south of Dublin, show remarkable examples of such oblique bedding as I have described.[2] Again at Howth, these beds, containing numerous marine shells which have been named by Dr. Scouler, may be observed

[1] The shells from the Wexford gravels were named by the late Prof. Edward Forbes, and considered by him and Sir C. Lyell to be of Pliocene age owing to the occurrence of *Fusus contrarius*, &c. But this evidence is inconclusive, and there can be little doubt the beds are the representatives of those here described.

[2] This section, accompanied by a wood-cut, is described in the Geol. Magazine, vol. viii. Here the Upper Boulder Clay, the Middle Sand and Gravel, and the Lower Boulder Clay may all be observed between the base of Killiney Hill and the Martello Tower of Bally-brack.

overlying the Lower Boulder Clay, and are ultimately surmounted by traces of the Upper Boulder Clay on the flanks of Howth Hill.

The Middle Gravels rise to considerable elevations on the flanks of the Dublin and Wicklow Mountains, where they also contain shells, most of which live in adjoining seas: they have been ably described in such positions by the late Mr. John Kelly,[1] and in greater detail by the Rev. Maxwell Close.[2] By these observers they have been found at elevations of 1,300 feet (Caldbeck Castle), and others not quite so high up, opposite Ballyedmonduff, on the road from Stepaside to Glencullen; and in the Killakee valley. The species from Ballyedmonduff have been determined by Mr. W. H. Baily, and are as follows:—*Trophon muricatus, Fusus* (part of columella), *Turritella communis, Ostrea edulis, Pecten* (two species), *Cardium edule, C. echinatum, Astarte compressa, A. elliptica, A. sulcata, Cyprina Islandica, Artemis lincta, Venus striatula, V. casina, Lutraria elliptica, Mactra stultorum, Tellina? Mya truncata, Pholas crispata, Balanus balanoides, Annelid* perforations. Mr. Close[3] observes that though all the above individual species live in the neighbouring seas, yet

[1] Journ. Geol. Soc. Dub., vol. vi. p. 133.
[2] Journ. Roy. Geol. Soc. Ireland, vol. iv. p. 36.
[3] 'On the more recent Geological Deposits in Ireland,' Journ. Geol. Soc. Dub., vol. iii. p. 61.

as a group they present a rather more boreal *facies* than those of the present coasts, and than those of the low gravels described by Professor T. Oldham ; but, on the other hand, a decidedly less northern *facies* than those from the Drift beds of Moel Tryfaen, in North Wales, which occur at 1,360 feet above the sea level.[1] Mr. Close supposes the gravels at Ballyedmonduff to have been carried thither by floating ice, and not to have been deposited where they occur, as they contain pebbles of limestone and other travelled stones. At the same time, re-collecting the elevation of those on Moel Tryfaen, which are probably representative and have all the appearance of having been formed *in situ*, and the prevalence of gravel beds at elevations but slightly below that just referred to, it seems probable that these beds have not been transported from any great distance.[2] Still, without some such agent as that suggested by Mr. Close, it is difficult to account for the presence of limestone pebbles.

In the north-east of Ireland, as pointed out by Professor Harkness, beds of stratified gravel and sand occur, with sea-shells belonging for the most

[1] First discovered by the late Mr. Joshua Trimmer, F.G.S.

[2] Mr. Close admits that these shell-bearing gravels belong to the division of the Drift series (middle sands and gravel) here described. They contain limestone pebbles which must have come from lower ground.

part to existing species, but indicating somewhat colder conditions than those which obtain at present.[1] In Co. Antrim near Ballycastle, these beds form conspicuous terraces rising on the flanks of the hills to an elevation of 600 feet. They rest on Lower Boulder Clay, and appear to be overlaid by an upper similar deposit.[2] A similar succession of beds occurs near Glenarm, as I am informed by Mr. Traill.

Besides the localities already mentioned, shells have been found in these gravels, in several localities inland, and at Muff on the hillside opposite Lough Foyle by General Portlock;[3] so that there can be no doubt the gravels were formed over the bed of the sea, which extended over the plain, and rose to very considerable elevations on the flanks of the mountains. In England they have their representatives in the shelly sands and gravels between the Upper and Lower Boulder Clays, bearing the same name, as originally proposed by the author when describing the Drift Deposits of Lancashire and Cheshire. They have been found at various eleva-

---

[1] Geol. Magazine, vol. vi. p. 542.
[2] A good section of these beds may be seen in the banks of the River Carey, four miles S. of Ballycastle.
[3] Gen. Portlock considers these shells to be of Pliocene age, but there can be little doubt they are referable to the interglacial period here described. Geol. Rep. Londonderry, &c. p. 165.

tions in these and the adjoining counties west of the Pennine Chain, climbing the flanks of these hills and of the Welsh Highlands. These facts lead us to infer a great depression of the land, extending over the northern portion of the British Islands, and gradually decreasing southwards. The general absence of erratic blocks, except such as have been washed out of the Lower Boulder Clay, indicates the disappearance of glacial conditions, such at least as prevailed during the preceding period; so that the deposits which were formed may properly be termed 'inter-glacial.' Assuming the greatest depression to have reached 1,500 feet below the existing level, the Irish area must have presented the appearance of an archipelago of islands; as the higher portions of the mountain groups could alone appear above the general level of the waters.

(3.) *Upper Boulder Clay.*—This deposit is sparingly distributed, as compared with the two preceding, as, owing to its position, it has suffered from denudation to a greater extent than they; nevertheless, its presence has been determined, particularly by Mr. E. T. Hardman, in several localities, both in the north and centre of the country. It is clearly displayed in the coast cliff, south of Killiney Hill, where it may be seen resting on an eroded surface of the 'Middle Sands and Gravels,' and sloping downwards towards the plain which opens out on Killiney Bay. Its presence is, however, much more extensive than is generally supposed, as it may easily be mistaken for the Lower Boulder Clay. This will be evident from the adjoining section at the marble quarries of Kilkenny, where it may be

FIG. 15.

Section of Drift Deposits at the Marble Quarry, Kilkenny.

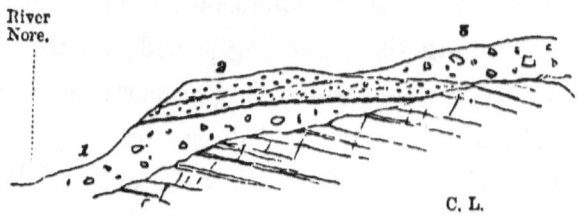

1. Lower Boulder Clay. 2. Middle Gravel.
3. Upper Boulder Clay. Length of Section about 50 yards.
C. L. Carboniferous Limestone.

seen resting on the Carboniferous Limestone, while the two lower members of the Post-Pliocene series are found cropping out on the banks of the River Nore.[1] (Fig. 15.)

The formation consists of reddish stiff clay with boulders, and bands of gravel and silt. It is sometimes a little sandy; in consequence of which, when the underlying gravels have some intermixture of

[1] This section has been given by Mr. James Geikie in the 2nd edit. of his excellent work, 'The Great Ice Age,' but is too important to be omitted here.

clay, the distinction has been often overlooked. To the practised eye, however, the distinction is sufficiently apparent, especially after the observer has studied the succession of the Drift deposits in the north of England. The great thickness to which this formation is capable of attaining is shown by the following section taken while sinking the shaft of the Modubeagh Colliery, belonging to the Leinster Colliery Company near Carlow :—

Section of Drift Deposits near Carlow.

Upper Boulder Clay—Stiff red clay with stones	84 feet
Laminated clay ('Book clay')	5 ,,
Middle Sands, &c.—Sand, sometimes clayey, with pebbles of limestone, &c.	25 ,,
Lower Boulder Clay—Strong clay with stones	8 ,,
Coal Measures—(Shales and fire-clay).	122 feet

Mr. Hardman has shown that the Upper Boulder Clay is well represented in the county Tyrone, capping the summits of the Drift hills where it has escaped denudation. He also shows that it occurs in numerous places in the Carlow district, resting on the middle gravels, and containing rounded and sub-angular blocks of rock, with polished and striated faces.[1]

It is exceedingly probable that in some parts of the country, especially in the west of Ireland, the

[1] Journ. Roy. Geol. Soc., vol. iv. p. 73; also Expl. Mem. Sheet 35 of the Geol. Survey Maps, p. 78.

Upper Boulder Clay has never been deposited, but though itself absent it is sometimes replaced, or represented, by large erratic blocks strewn over the upper surface of the interglacial gravels.

Such blocks, consisting of large slabs of Carboniferous grit, are strewn over the surface near the village of Killeely S. of Swinford in Co. Mayo. So numerous are these in some places that they might be mistaken for the outcroppings of the solid rock, were it not for sections in pits and banks of streams which reveal thick beds of gravel lying underneath.

At other times, the clay of the formation has been washed away; but the large boulders which it once contained are left strewn on the surface of the subordinate beds of gravel. Thus in the valley of the Yellow River which flows through the range of the Ox Mountains in Sligo, the stratified sands and gravels form terraces and ridges along the sides of the valley, and are generally strewn with large boulders of local, or foreign, rocks, such as Carboniferous grit from the south. In other places, however, these blocks are seen to be imbedded in reddish clay resting on very coarse gravel, which I consider to be the representative of the Upper Boulder Clay near its westerly limit.[1]

---

[1] I am not aware of any instances of this deposit west of the Ox Mountains.

Upper Boulder Clay.

It will be evident from the above description that the Upper Boulder Clay is due to the recurrence of glacial conditions in some sort, but probably not to anything like the extent, or exactly of the same kind, as existed during the period of the Lower Boulder Clay, or Till. We have seen that the phenomena connected with this latter deposit can only be fully accounted for by the supposition of an ice-sheet spreading itself over the land. But in the case of the newer deposit, there is great difficulty in admitting this view, as a second ice-sheet would probably have ploughed out all the older deposits of the Post-Pliocene age, and left them confusedly piled up against the flanks of the mountains, or even carried out to sea. But the Upper Boulder Clay generally rests on the interglacial beds in such a manner as not to show evidence of forcible attrition. The surface of the gravels is often eroded as by water action, but seldom displaced as by a solid mass of ice.[1] In the north of England the Upper Boulder Clay is sometimes stratified. In this country, as noticed by Mr. Hardman, it also sometimes shows traces of stratification. It is, in fact, just such a deposit as might be supposed to have been formed under the sea, the waters of

---

[1] In some cases, however, the interglacial gravels show evidence of displacement, as in the neighbourhood of Balbriggan, but it is probable (as suggested by the late Prof. Jukes) this is due to the stranding of floating icebergs.

which were laden with ice-rafts and bergs bearing stones and boulders which they would constantly deposit as they melted away, and which were themselves rendered turgid by streams of glacier-water entering from various directions. It is in this manner, as I conceive, the Upper Boulder Clay was really formed. A similar deposit is probably now in course of formation over the floor of the Greenland Sea, as far south as the banks of Newfoundland, due to conditions such as those above described. If this be so, it is clear we must assume a considerable elevation of the land (corresponding to a shallowing of the sea-bed) at the close of the period during which the inter-glacial gravels were formed. Such an elevation would be necessary in order to allow of mountain groups of sufficient extent and elevation for the formation of glaciers in their valleys, and in order to become centres of dispersion, from which bergs and rafts of ice could float off and melt in the waters of the surrounding sea.

It is not, perhaps, difficult to restore the physical features of our country under such conditions. I am unable to state the actual, or original, limits of the Upper Boulder Clay above the sea, but if we assume from 800 to 1,000 feet, we shall probably not be far from the truth. If such be the case, all the hills and mountains above the level of 1,000 feet must have formed groups of islands[1] from which icebergs and rafts were dispersed over the surrounding waters; the Donegal and Derry Mountains on the north, those of Mayo and Galway on the west, those of Kerry, Cork, Tipperary, and Waterford on the south-west, those of Wicklow and Dublin on the south-east, and those of Down and Armagh on the north-east appear to have formed an archipelago of snow-clad isles, represented in the British area by similar groups.[2]

[1] A map representing the position of land and sea of the Irish area during the period of greatest depression will be found in Lyell's 'Antiquity of Man,' 4th edit., pp. 325-8. (Fig. 42.)
[2] If the depression along the flanks of the Wicklow Mountains during the stage of the Middle Gravels amounted to 1,500 feet (as measured by the existing sea margin) and that during the stage of the Upper Boulder Clay for 1,000 feet, there would have been a rise of 500 feet in the sea-bed between these stages.

CHAPTER V.

POST-GLACIAL DEPOSITS.

Post-Glacial formations, Mountain Terraces, Eskers and Local Moraines.—With the close of the formation of the Upper Boulder Clay the history of the Glacial Epoch properly ends. At its close there was a gradual elevation of the land from beneath the waters, accompanied probably by pauses; the climate became milder, and the glaciers retreated, little by little, up the mountain glens, until they disappeared altogether under the genial influences of a warmer sun and less arctic temperature. Into the causes of this change it is not necessary for me to enter here, the subject having been sufficiently handled by others better qualified than myself to deal with it, and to their writings I must refer the reader.[1] The physical evidence of a colder climate preceding the existing temperate one in these latitudes obliges us

---

[1] Lyell's 'Principles of Geology,' Croll's 'Climate and Time,' James Geikie's 'Great Ice Age,' will probably be found to contain sufficient information to satisfy the enquiring reader.

to infer an intermediate Post-Glacial period, during which the change from the one to the other, and from the former conditions of sea and land to the present, must have been gradually progressing. On the whole, the land was gaining on the sea; and as lower slopes amongst the mountains successively rose into the air, and were exposed to wave action, terraces composed of the recently stratified materials would naturally be formed whenever a pause took place in the upward movement.

Such terraces are not uncommon amongst the Irish mountain groups. One of them may be very clearly observed, when looking up the valley of the Kilkeel river from the south, skirting the base of Slieve Lough Shanagh in the Mourne Mountains at an elevation of about 1,000 feet above the sea. Amongst the mountains of the west of Ireland similar terraces are of frequent occurrence, and can often be better seen at some little distance than when standing immediately on their surfaces; thus two distinct terraces may be observed on the flanks of the mountains running up from Killary Harbour to Delphi when they are looked at from the south bank of the harbour, but are not so conspicuous when the hill-side itself is examined, the observer not being then in a favourable position. Not far, however, from this spot, and facing the head of the harbour above Leenane, there are two very well-formed terraces of gravel: one at an elevation in the upper surface of about 60 feet, and the other at an elevation of about 200 feet; this latter is very extensive, and is traversed by the road to Cong. Similar terraces, consisting of slightly sloping upper surfaces and abrupt margins, and formed for the most part of Drift or Moraine materials, may be observed at the base of the quartzite mountains near Recess in Connemara, and along the Lough Inagh valley.

Mr. Kinahan records similar 'well marked terraces which appear to be ancient sea-margins,' on the flanks of Slieve Aughta in the neighbourhood of Lough Graney, the highest being at about 1,200 feet, and the lowest a little above 300 feet.[1] In the Burren country, Co. Clare, there are remarkable terraces formed amongst the nearly level beds of Carboniferous Limestone; but it is doubtful to what extent they can be considered ancient sea-margins. It is very probable, however, that some of the incipient contours may have been determined during the rising of the land.

Eskers.—While the land was still being elevated, and fresh tracts were emerging into day, or were

[1] 'Notes on the Drift of Ireland,' Journ. Roy. Geol. Soc., vol. i. p. 198.

Eskers.

being brought within the reach of surface waters, it may easily be imagined that the tidal and other currents, being forced to oscillate within narrow channels bounded by the ridges of the unsubmerged land, would act with considerable effect on the soft materials of the Drift, both in sweeping them away, and in piling them up along tortuous lines in the form of embankments. And such may the Eskers be regarded. They consist of long mounds or banks of gravel, formed for the most part out of the inter-glacial gravels, often running for miles, and assuming directions depending on those of the adjoining hills; they are confined to the plains, and their upper surfaces are frequently strewn with large erratic blocks, probably derived from the Upper Boulder Clay, which had itself been washed away during the emergence of the land. These Eskers are found at intervals over the great central plain, and as far north as the valley of the Lagan on the borders of Down and Antrim.[1]

Mr. Kinahan observes that the Eskers in the strip of country lying between Dublin and Galway form a compound bar, consisting of well-defined ridges or 'bar-eskers,' and in other places of shoal-eskers. The *bar-eskers* from Galway to Tullamore, or thereabouts,

---

[1] Some of the Eskers have been represented on the maps of the Geological Survey, and described in the Explanatory Memoirs.

are usually on ground under the 250-feet contour line, and from Tullamore to Dublin on ground under the 300-feet contour; while the *shoal-eskers* towards the west are on ground between the 250-feet and the 300-feet contour, and towards the east between the 300-feet and 400-feet contours. From these facts Mr. Kinahan infers that the land was then between 300 and 400 feet lower than at present, having since risen more to the east than it did to the west.[1]

Mr. Kinahan considers that icebergs floated about in the Esker-sea, carrying blocks of rock, and strewing them over the sea bed and on the Esker-banks.[2] In this way he accounts for the remarkable stream of erratic blocks of porphyritic granite, with large pink-coloured crystals of felspar, which he and Mr. O'Kelly[3] have traced from the neighbourhood of Oughterard, eastward from Galway Bay by Loughrea and Ballinasloe to the flank of Slieve Bloom. But, as I have already shown, the sea was laden with bergs and rafts of ice carrying blocks of rock during the period of the Upper Boulder Clay; and it seems to me much more probable that such trains of erratics as those described by Mr. Kinahan,

[1] 'On the Drift of Ireland.' *Supra cit.*, p. 200.
[2] *Ibid.*, p. 202.
[3] Explanation, sheet 127 of the Maps of the Geol. Survey, p. 26.

together with similar ones in other districts, are the relics and monuments of this epoch; somewhat as the large slabs of grit which are strewn over the Chalk Downs of Berks and Wilts are the relics and monuments of the Lower Tertiary beds with which they were once associated, these beds themselves having disappeared.

The Rev. M. Close, in his admirable paper on 'the General Glaciation of Ireland,' gives a summary of the observations of such erratic blocks derived from various sources, which shows how extensive has been the dispersion of such blocks by the agency of floating ice; and to this paper I must refer my reader for fuller details than can be given here.[1]

Local Moraines.—When describing the conditions under which the Upper Boulder Clay was formed, I showed that we had reason for believing that the mountains of Ireland rose out of the surrounding sea in the form of groups of snow-clad islands, down the valleys of which glaciers descended into the sea, giving birth to icebergs. During the progress of elevation of the land, and the replacement of the glacial, by more temperate, conditions of climate, we may suppose that these snowfields gradually melted away, and that the glaciers which were fed by them withdrew step by step up the valleys. This retreat

[1] Journ. Roy. Geol. Soc. Ireland, vol. i. p. 228.

of the glaciers was doubtless a slow and lengthened process marked by pauses, during which moraines would be formed along their sides and at their lower extremities. Such moraines are strikingly exhibited amongst the mountains of North Wales and of Cumberland and Westmoreland, and are also to be noticed occasionally, though less conspicuously, amongst some of the mountain districts of Ireland. I was unable to identify them amongst the mountains of Mourne, as it appeared to me, when examining this fine group of granitic elevations, that the masses of moraine-like matter to be found on the slopes and amongst the valleys resolved themselves into sheets, or beds, descending into, and connected with, the Lower Boulder Clay. On the other hand, local moraines accompanied by local striations may be observed amongst the mountains of Wicklow, Kerry, Waterford, and Galway. The cross-striations on the glaciated surfaces of the rocks in the vicinity of these mountains are referable to two or more systems of ice movement :—one of these to the earlier and more general glaciation of the Lower Boulder Clay stage, and the others to the more recent action of local glaciers originating in the mountains themselves. They have been observed by Mr. Kinahan in Mayo, Galway, and Clare; by Mr. Foot in South Leitrim; by Mr. Symes on the shores of Killala Bay; by Mr. Close in Co. Wicklow; by Messrs. Jukes and Du Noyer amongst the Commeragh Mountains in Waterford, and the mountainous promontories of Kerry. These local striations require to be carefully distinguished from those of a more general range and earlier date; and to be fully described would require a separate treatise, together with a glacial map, for each individual mountain group. The local moraines generally consist of mounds, hummocks, and banks of sandy clay, containing stones and subangular blocks of local rocks confusedly heaped together: the blocks are sometimes glaciated on their surfaces, and in various directions. These local moraines, when thrown across a valley, as 'terminal moraines,' help to form lakes, by damming up the waters of the streams which enter it above. At other times, however, the river has cut a channel for itself, generally near the end of the moraine where it originally came in contact with the side of the bounding ridge, and thus, the waters having been drawn off, the lake has disappeared.

When the moraines are lateral they are piled up along the flanks of the valleys, and are less conspicuous than those above described.

It would be impossible to do more than give a few out of the numerous examples of local moraines to be found amongst most of the mountains of Ireland; but the following may be mentioned:

The remarkably straight and picturesque valley of Glenmalure, which lies along the line of a large fault and is drained by the Avonbeg River, furnishes at least two examples of terminal moraines. On ascending the valley from 'the Vale of Ovoca' through Ballinacor Park, we are struck by the large number of huge boulders of granite which have been brought down from the interior of the mountains. One of these near the road measures $12 \times 7 \times 4\frac{1}{2}$ feet, and from this a large fragment had apparently been broken off and lies alongside. At the upper end of the park near Strand Bridge, and where a lateral valley enters from the south, immense piles of moraine matter laden with granite boulders lie across the valley, extending for some distance laterally along the northern side, and cut through by the Avonbeg near its centre. Here we have clearly an old terminal moraine of the glacier which formerly extended down this noble glen, and drained the snow-fields of Lugnaquilla and the neighbouring heights. Above the moraine, the flanks of the valley may be observed to be glaciated to a height of about 500 feet above the bed of the river; above which, traces of glaciation become indistinct, or entirely disappear.

A second and smaller moraine occurs about two miles higher up Glenmalure, near the hotel, in the form of an irregular embankment which has evidently extended originally across the valley from side to side. The terraced surface of the valley above the moraine may once have been the bed of a lake which has since been drained, the river having cut a deep channel for itself through the moraine. This is probably one of the latest examples of local moraines amongst the Wicklow Mountains; formed during a pause in the retreat of the glacier towards the head of the valley, while the snows of the surrounding heights were melting away.

The bank which is thrown across the lower end of Lough Bray in Co. Wicklow is considered by Mr. Close to be a terminal moraine.

Leaving the Wicklow Mountains and coming to those of West Galway, there is, amongst others, an example of a terminal moraine at the entrance to Glen Inagh where it opens out on the wide Glendalough Valley in West Galway, a favourite position for such accumulations. The chain of loughs stretching along the southern slopes of the Connemara Mountains affords numerous instances of rock-basins and moraines, while the rocks of schist and crystalline limestone which rise out of their waters are strikingly ice-worn. The local glaciation has here obliterated the earlier general glaciation; so that east of the meridian of Lough Oorid the ice has moved eastward, and to the west of the same line, in an opposite direction. The great tract of undulating moorland which stretches southwards to the shores of the Atlantic is dotted over with loughs which are either rock-basins scooped out by the glacial ice, or pent-up reservoirs where moraine matter has been left by the retreating ice. Thus the picturesque little lough of Glendalough at Recess has been pent up at its western end by a great mound of moraine matter, which has been thrown across the original valley, and by which it is separated from L. Nacoogarrow. Immense masses of moraine matter are strewn along the shores of Lough Inagh and the slopes of Derryclare, probably left behind as the glacier dried up, and disappeared.

Along the southern shores of Killary Harbour, that remarkable fiord which penetrates the Western Highlands for a distance of about twelve miles from the Atlantic, the glacial phenomena are very striking. The rocks are intensely glaciated, and scored with groovings pointing down the valley, while masses of moraine matter with huge boulders are strewn along the shore.

CHAPTER VI.

RECENT DEPOSITS.

Raised Beaches and River Terraces.—Though our coast affords occasional evidence of local depressions,[1] the features which most strike us as bearing on the question of oscillations of level are the raised beaches, occurring in the form of terraces, rising above the reach of the highest tides, and often bounded inland by old coast cliffs. Traces of such terraces may be observed at the head of Killary Harbour, and along the base of the quartzite hills of Connemara; along the shore of Kenmare and Glengarriff Bays; at Ardmore Point on the coast of Wicklow, and other localities. But the most striking and continuous of all the raised beaches I am acquainted with is one which is traceable at intervals along the northern and eastern coasts of Ireland, and which is undoubtedly the representative of the 'twenty-five feet terrace' of the western coast of Scotland, with the peculiarities of

[1] See Mr. Harte on the Physical Features of Donegal, *supra cit.*, p. 26.

which geologists have become familiar by the writings of the late Mr. Smith of Jordan Hill, and, more recently, those of Professor Geikie. This raised beach forms a fringe of nearly level gravelly soil skirting the coast of Kintyre, Arran, Rothsay, and the sinuous shore of the Clyde; while inland it is generally bounded by a cliff, or steep slope sometimes perforated by caves, the bottoms of which are strewn with rounded boulders, just as they were left at the recession of the tide thousands of years ago. Crossing to the Antrim coast, we soon recognise similar peculiar features in the protected nooks and bays of this bold sea-margin. The terrace may not be quite so elevated above the high-water line as on the opposite shore, the average elevation being about fifteen feet, and the sea waves have in some places made great inroads upon its margin, but it is not the less determinate in certain localities. Commencing at the north we shall follow the course of the beach at intervals along the shore.

The coast of Inishowen, in Co. Donegal, sometimes has a trace of this beach in the form of a terrace in the less exposed situations. I have noticed it at Culmore and Culdaff, rising about fifteen feet above the highest tides. On the southern shore of Lough Foyle around Bellarena Station, the raised beach extends over a wide area, the ancient sea-margin being indicated by an abrupt bank of drift, or harder rock, lining the terrace inland. Near Castlerock, the upper surface of the terrace is covered by huge sand dunes, piled up by the westerly winds; and to the west of the terrace, considerable tracts of rich alluvial land have been reclaimed from the sea by the aid of embankments, resembling the coasts of Holland, or of the Lincolnshire Fens; and bear evidence of the energy and enterprise of the farmers of this part of the country.

Fig. 16.

Old sea-stack of basalt. Island Magee.

At Portrush it occurs as a shelly gravel resting on the dolerite and the indurated Liassic beds of that district. But it is along the eastern coast of Co. Antrim that its features are most marked. That the coast has here been raised is evinced, not only by the presence of the narrow marginal terrace, but by the old coast cliffs perforated by caves now well out of reach of the waves. Sea-stacks of rock, standing erect as they were left when placed out of reach of the breakers by which they had been dissevered from the interior solid masses, are conspicuous monuments. One of these stands above the harbour of Ballycastle, and another is represented in fig. 16, from the E. coast of Island Magee. The caves may be seen penetrating New Red Sandstone at Red Bay and Glenarm, and basalt along the coast N. of Larne. Three of them at Port Ballintoy were explored some years ago by Mr. James Bryce [1] (the late Dr. Bryce of Glasgow), in company with Dr. McDonnell, and yielded bones of horse, ox, deer, sheep, goat (?), badger, otter, water-rat, and several birds. Another cave on the west side of Carrickareede Island yielded similar remains.

The best section of the raised beach-bed, perhaps, to be obtained on the Antrim coast is that which is laid open in some gravel pits near Larne Harbour. The old sea-bed is here elevated into a terrace about fifteen to twenty feet above H. W. line, and is seen to be composed of stratified and waterworn gravel, with numerous blanched marine shells, and (what is still more remarkable) with flint-flakes of human workmanship. The shells belong to species inhabiting the neighbouring sea, and the worked flints are of that rude form and finish known as 'Palæolithic.'

[1] Trans. Brit. Association, 1834, p. 658.

Another excellent section may be observed at Kilroot on the north shore of Belfast Lough, where the shells and worked flints are exceedingly abundant. It is here that these works of ancient art were first discovered by members of the Belfast Naturalists' Club; they were afterwards described by the late Mr. Du Noyer, who, judging by the great number of the chips of flint accompanying the arrowheads, or spear-heads, came to the conclusion that the shore of Kilroot had been an ancient Palæolithic workshop, where weapons of war, or of the chase, were made from the chalk-flints of the adjoining hills. The following species of shells have been identified by the Rev. Dr. Grainger from the raised beach at Larne Curran: *Anomia ephippium, Ostrea edulis, Pecten varius, Cardium edule, Kellia suborbicularis, Lucina borealis, Tapes pallustra, Tellina tenuis, Corbula gibba, Saxicava rugosa, Patella vulgaris, Trochus magus, Turritella terebra, Littorina obtusa, Purpura lapillus, &c.*[1]

Extending our observations further south, we occasionally find traces of the beach in the form of a narrow terrace from twelve to fifteen feet above high-water mark along the Downshire coast, as at Killough Bay near Ardglass (see fig. 14), and in Dundrum Bay. They may also be observed at Greenore, form-

[1] Trans. Brit. Assoc. (Belfast meeting, 1874), p. 73.

ing a terrace of shelly gravel about ten feet above the sea at the entrance to Carlingford Bay. Under the town and neighbourhood of Dundalk the terrace is rather widely spread, formed of shelly gravel. It may also be noticed at intervals along the coast southwards; as for example at Lowther Lodge near Balbriggan,[1] and towards Dublin Bay, where it almost descends into the sea, the elevation being only five or six feet at Clontarf, and along the narrow neck which joins the Hill of Howth to the mainland. In Dublin Bay, it merges into the old estuary of the River Liffey, forming the level terrace on which the Custom House, the Bank of Ireland, the University, and Sackville Street are built.

It seems to me that this terrace subsides into the general level of the coast-line in Dublin Bay. It may possibly be represented by the terrace at Bray, and the remarkable shingle beach, partly piled up by the waves and current, which stretches for several miles along the coast north of the town of Wicklow. The elevation here is, however, only slightly above that of the high-water line; and comparing this with the level along the coast of Antrim, it is clear that the

---

[1] Here the following shells occur:—*Litorina litorius, Pecten maximus, Purpura lapillus, Cardium echinatum, C. edule, Rostellaria pes-pelicani, Mya arenaria, Patella vulgaris, Trochus. Dentalium. Turritella communis, Turbo cinereus.*

surface of the terrace has a general slope downwards from the north to the centre of Ireland.

The 15-feet beach of the north-eastern coast may be considered as the most recent representative we can point to of coast and land elevation. Professor Geikie has shown that there are some grounds for believing the 25-feet beach of Scotland (the representative of that here described) has been elevated into land since the Roman occupation of Britain. Whether this is the case or not, it is certain from the occurrence of buried canoes in this terrace in the Clyde valley, and the presence of the worked flints, associated with the shells in the stratified gravels at Larne and Kilroot, that the coast has been raised since the occupation of the British Islands by ancient Celtic tribes. This brings the history of physical changes as far down towards our own times as it is necessary for us to follow it. It is at this point that the archæologist and historian takes up the pen which the geologist lays aside.

River Terraces.—In connection with the subject of the elevation of the coast and raised beaches, we are naturally led on to refer to the formation of river terraces. These may be observed along the sides of the principal rivers of our island, where they approach the coasts and enter upon the last stage of their course, which is generally smooth and tranquil. The terraces may be observed forming almost level surfaces bounded by abrupt descents which lead down to the alluvial flat still liable to floods. Where sections are exposed, it will be found that, like the alluvial flats themselves, the terraces are composed of river gravel, generally overlaid by a thin stratum of silt; and we therefore infer that they are themselves old alluvial flats which were once occupied by the waters of the river, but which are now forsaken, the stream having deepened its channel. Sometimes two or more of these terraces may be observed, indicating successive deepenings.

Now, such changes in the river bed must evidently follow the rising of the coast; for where the coast rises the rivers are converted into rapids, or waterfalls, which commence to cut back their channels inland, until a balance between the eroding power and the inclination of the river bed is established. Meanwhile, the channel being lowered, the former alluvial flats are laid dry, and new ones in a lower position are constructed. The old terraces which are so conspicuous along the valley of the River Boyne, in the neighbourhood of Navan and Slane, may therefore be traced back to the period when the sea coast itself was elevated, as shown by its raised beach; and on the opposite coast of Ireland, the remarkable terraces of the Errif Valley, above Killary Harbour, may have some connection with the old sea terraces or beaches which have already been described as occurring in that neighbourhood. I have already referred to the old terrace along the banks of the Liffey at Dublin, which merges into the slightly raised beach of this part of the Irish coast.[1]

Amongst the river valleys of Wicklow, numerous instances may be observed of old river terraces; as for example, at the lower end of the vale of Glenmalure above the junction of the rivers Avonbeg and Avonmore. But of these the most remarkable and interesting, from its historical and architectural associations, is the terrace at the lower end of Glendalough, at its confluence with the vale of Glendasan, upon which stands the Round Tower and the Church of St. Kevin. This terrace is composed of stratified gravel of rounded pebbles and sand banked up against an old moraine which has been thrown across the valley. The upper surface is level, and it rises about 20 feet above the bed of the Glenealo River. As the site for the interesting ecclesiastical buildings by which it is surmounted, it was admirably chosen. After the ruins of Iona which Dr. Johnson has immortalised, perhaps there are none in the British Islands which impress the visitor with more interest than those of 'the Seven Churches' rising in

[1] Sse Geol. Survey Map, Sheet 112.

chaste simplicity amidst the solitudes of these deep valleys; monuments of the faith and piety of a past age.[1]

[1] A very interesting little handbook and guide to the ruins of Glendalough, with a historical sketch, has been written by Mr. Joseph Nolan, M.R.I.A. of the Geological Survey of Ireland.

PART II.

PHYSICAL GEOGRAPHY OF IRELAND.

CHAPTER I.

MOUNTAINS.

When a traveller explores a new country, the objects which are calculated to arrest his attention are, first, the mountains, then the plains and great valleys, and ultimately the rivers and lakes; and in the following attempt to elucidate the origin of the physical features of Ireland, I shall adopt the order here indicated, as it seems to correspond not only with the usual process of observation, but with the order in which the physical features themselves were developed. The origin and geological ages of the respective mountain chains, or groups, first demand our consideration. But it is desirable that I should state at the outset what I mean by 'the geological age' of a group of mountains; for, as we shall see, some of the Irish mountains have been either submerged beneath the ocean, or partially buried beneath newer strata after they had been in existence many ages before. Thus it happens that the Highlands of Scotland have at one time been deeply buried beneath strata of Devonian and Carboniferous ages. The mountains of North Wales have been, in part at least, entombed in Carboniferous, Triassic—possibly Jurassic and Cretaceous—beds; even the central core of the Alps has more than once almost disappeared beneath enormous masses of Secondary and Tertiary formations. But after all these vicissitudes the mountains have emerged again into day; changed, indeed, in form and feature, but ever grand and venerable; the noblest objects on which man can gaze, and made use of by him as emblems of the majesty of the Great Creator Himself. After various vicissitudes and periods of obscurity, they now stand before us in all the beauty and loftiness of maturity, and in all their wonderful variety of form, shadow, and colouring, and as we gaze upon them our thoughts find expression only in such words as those of England's great epic poet:[1]

> These are thy glorious works, Parent of Good!
> Thyself how glorious then!

---

[1] Still grander is the language of the Psalmist of Israel when he points to the mountains as emblematical of the eternity of

The Birthday of Mountains.

When we speak, therefore, of the age of any of the Irish mountains, it must be understood that we mean their actual birthday (so to speak) and not any subsequent *renaissance*, or redevelopment after a period of obscurity. Now, there is only one way in which such an epoch can be determined in the case of any particular group of mountains; namely, by observing the relations of the strata of which the group is composed to the newer formations with which it comes in contact, and determining the geological age of each. Thus, if we find a mountain group composed of strata belonging to the Lower Silurian system (as indicated by their fossils or otherwise), to be overlaid along its base *unconformably* [1] by strata of the age of the Old Red Sandstone, we have here evidence that the beds forming the mountain group have been forced out of their originally nearly horizontal position, and subjected to denuding action before the newer beds were deposited over them. Now, as denudation can only take place at the surface of the ocean, or under the atmosphere itself, we are led to conclude that the denuded beds had, in such an instance, been converted into a land-

---

Jehovah: 'Before the mountains were brought forth, or ever the earth was made, from everlasting to everlasting Thou art God.' Ps. xc. 2.

[1] That is, when the strata of the newer formation rest on the disturbed and denuded edges of those of the older. See p. 26.

surface before the epoch of the Old Red Sandstone;—and this was, in geological language, *the date of their birth*.

Upon the principles above laid down, we shall now proceed to determine (as far as the evidence admits, for it is not always conclusive), what may be the respective ages of the different mountain groups of Ireland, and the special features connected with their development. The description already given of the formations themselves, and of their mutual relations, will have prepared the reader for what is to follow.

Before, however, entering upon an account of the geological age of the mountains of Ireland, it will be desirable for me to endeavour to sketch out the leading physical features of the country, as represented by its mountains and valleys.

Ireland may be described as an island consisting of a great central plain, bounded in various directions near the coast, but not entirely surrounded, by groups of mountains. Along the west, south, and north the coast is deeply indented, and the central plain is diversified by numerous lakes, and sluggish rivers which find their way out to sea at the head of the bays and estuaries. A line drawn across the centre of the country from Dublin or Dundalk Bay on the east, to Galway Bay on the west, will meet with no higher elevation than that of about 250 feet above the sea; but a section in every other direction will be found to cross a mountainous ridge bounding at each extremity the central undulating plain. This plain is formed of Carboniferous limestone, which occurs in beds not much removed from the horizontal position; except in the neighbourhood of local disturbances, or of several eminences which rise above the general level, towards the southern portion of the country, and which are composed of older formations.

For the purposes of classification, the mountain groups may be divided into the following:—(1) The North-western Highlands of Donegal and Derry. (2) The Western Highlands of Mayo and Galway including Connemara. (3) The South-western Highlands of Kerry and Cork, with the outlying elevations of Mauherslieve, the Devil's Bit, Slievenaman, Knockmealdown and Galtymore, which all physically belong to this group. (4) The South-eastern Highlands of Wicklow and Dublin; and (5) the North-eastern Highlands of Mourne, Carlingford, and Slieve Gullion. Besides these, there are of course many minor hills, variously composed, and of different ages, which are not of sufficient magnitude to entitle them to rank as mountains;—though it must be confessed that it is often difficult to say where the term 'hill' should be substituted for that of 'mountain.'

I have already given some account of the geological structure of these mountain groups, when describing the formations of which they are composed, and our principal object is now to endeavour to ascertain their actual and relative geological ages, bearing in mind those principles which I have laid down for our guidance in this inquiry.

CHAPTER II.

NORTH-WESTERN AND WESTERN HIGHLANDS.

In searching amongst the groups I have named above for the most ancient, we naturally refer to the formation of greatest antiquity, and we find ourselves amongst the north-western, the western, and the south-eastern Highlands, formed, as I have already shown (p. 10), of Lower Silurian beds generally converted into crystalline schists, quartzites, and gneiss by that deepseated hydro-thermal process which we call 'metamorphism.' I have already shown that the metamorphic rocks of Donegal, Mayo, and Galway belong to one great geological system continuous with that of the central Highlands of Scotland, and that the epoch at which this transformation of the original strata took place is accurately determined for us by the position of the Upper Silurian rocks on both sides of Killary Harbour, and in the direction of Lough Mask.[1] Here it is that the Upper Llandovery

---

[1] Some account of the physical features of these districts has already been given at p. 22, and need not therefore to be repeated.

Beds, consisting of red and grey shales, grits, and conglomerates, are found resting discordantly on the eroded edges of the metamorphic rocks, the former being well stored with fossils by which their age has been determined, and often containing rounded pebbles and blocks of quartzite, schist, and gneiss derived from the metamorphised Lower Silurian beds. These newer beds rise in a series of ledges and terraces into the summit of Muilrea, 2,688 feet, and form a range of hills extending to the shores of Lough Mask, including the remarkable table-land of Slieve Partry, or 'Joyce's Country.'

From the position of these beds with reference to those upon which they rest, it is evident that they have not shared in the metamorphism to which the older Silurian rocks have been subjected. The metamorphism is, therefore, of older date than the Upper Silurian, and belongs to a long unrepresented period intervening between the Lower Silurian on the one hand, and the Upper Silurian on the other.

I have dwelt somewhat at length upon this point because of its importance. The period of metamorphism was accompanied by great disturbances of the Lower Silurian strata; they were bent into folds and flexures, elevated out of the waters of the sea in which they had been deposited, and subjected to the operation of denuding agents—the sea waves, the rains, and rivers of the period,—which swept away large masses of the more exposed portions, hollowing out valleys, and thus in reality creating the mountains. One of these hollows lay along the line of the Upper Silurian hills I have just described. On the resubmergence of the land the sides of this valley were subjected to wave action, and shingle beaches, now occurring in the form of conglomerate beds, were strewn over the floor of the sea. Here, then, we have the first indications of mountains;—that is, of old land surfaces, raised, we know not how high, into the air, and bounded by valleys which ultimately became arms of the sea, and were filled up as the land subsided by beds of shingle, sand, and clay derived from the waste of the neighbouring lands. It has also been shown that the rock masses of the Donegal Highlands are of similar age and composition to those of the West Galway and Mayo Highlands, and that they have been subjected to similar changes and at the same period of geological time; so that from all these considerations we arrive at the conclusion that the epoch of the formation of the north-west and western Highlands is 'Pre-Llandovery,' to use a technical phrase; in other words, an epoch immediately preceding the formation of the Upper Silurian series, the base of which is formed of Upper Llandovery Beds. Such is the age of our oldest mountains.

CHAPTER III.

SOUTH-EASTERN HIGHLANDS.

THE range of mountains extending from the shores of Dublin Bay through Wicklow into the valley of the River Barrow, in Carlow, is formed of granite, which penetrates the Lower Silurian beds in the manner already described (p. 12); these latter being considerably altered, or metamorphosed, in proximity to the granite.[1] It is remarkable that although the Cambrian rocks occupy a large tract of country along the sea coast, the granite never penetrates these older rocks, but is limited to the Silurian area.

Along the central line of the granite, the mountains attain their highest elevations, rising by long sweeping moorlands often, however, bounded by precipitous or steep cliffs and escarpments, into the culminating points of Kippure (2,473 feet), Duff Hill (2,364 feet), Douce Hill (2,384 feet), and Lugnaquilla (3,039 feet). These mountains give

---

[1] This is indicated by the development of mica along planes of foliation.

birth to numerous fine streams, such as the Liffey, the Vartry, the Avonmore and Avonbeg, which uniting at the well-known 'Meeting of the Waters,' forms the river Ovoca; and being still further augmented by the confluence with the River Aughrim, enters the sea at Arklow. The southern portion of the range is drained by the River Slaney.

These rivers traverse tracts of Silurian and Cambrian rocks, along valleys remarkable for their depth; while to the naturally varied character and ruggedness of the landscape there has been imparted an aspect of richness and softness owing to the extensive woods with which the banks of the valleys are clothed. Some of these river valleys are very ancient, dating probably from the Carboniferous period, or even earlier:[1] to these I shall have occasion to refer again, and now pass on to consider the matter more immediately in hand—namely, the date at which the mountains of this region themselves were 'brought forth.' In order to do this we must assume that the birth of the mountains corresponds with the intrusion of the granite. Though this rock is probably in part metamorphic, it is also in part intrusive; and its formation was (we may suppose) accompanied by a considerable elevation of the Silurian rocks, forming a ridge

[1] See Mr. Kinahan's paper 'On the Estuary of the River Slaney.' Journ. Roy. Geol. Soc., vol. iv. p. 60.

which maintains its preeminence, though subjected to extensive denudation, down to the present day.

Now, we have no very precise information regarding the age of the Leinster Granite further than this: that its formation was antecedent to the period of the Old Red Sandstone. This is clear from the position of this latter formation on both sides of the River Slaney above New Ross, where the Old Red Sandstone, and then the Carboniferous Limestone, gradually overlap the Lower Silurian beds, and finally rest on the eroded surface of the granite itself near Bagnalstown.[1] The granite, and as we may infer, the mountains which are formed of it, is therefore older than the epoch of the Old Red Sandstone; but having reached this point in our inquiry, our evidence ends as regards direct superposition of strata; and perhaps it would be wise not to venture on indirect evidence in order still more closely to determine the date of irruption.

In looking at a geological map, however, we cannot close our eyes to the fact that the granite is protruded along a certain definite direction. This general direction is ENE. and WSW.; and it will be observed that it corresponds nearly with the out-

[1] That the granite was consolidated before the Carboniferous period we have abundant evidence in the neighbourhood of Dublin, where pebbles of granite have been found imbedded in the limestone.

bursts of granite and trap in Donegal, the date of which we have ascertained to be immediately preceding the formation of the Upper Silurian beds. I cannot help attaching importance to this parallelism of direction, because it is intimately connected with parallelism in the action of terrestrial forces, which, from the days of Elie de Beaumont, is admitted by physicists to be some evidence of contemporaneity.[1] I do not insist very strongly upon the application of this principle in the present instance, and will only go so far as to say that the granite mountains of Wicklow are certainly older than the period of the Old Red Sandstone (or Devonian), and probably older than that of the Upper Silurian.[2]

We now pass on to the consideration of the age and origin of the remaining groups, amongst which we shall be brought into contact with natural operations of a different character from those we have been hitherto considering.

[1] M. de Beaumont may have carried his theory too far, but it certainly contains a germ of truth.

[2] And therefore synchronous with those of Donegal, Mayo, and Galway.

CHAPTER IV.

SOUTH-WESTERN HIGHLANDS.

Mountains of Kerry, Cork, and Waterford.—The ranges of hills and mountains I have hitherto described are, as we have seen, connected intimately with igneous, or hydrothermal, action resulting in metamorphism of the strata; but in the case of the ranges now to be considered no such action is observable. Here we are brought face to face with the results of mechanical forces arising from the contraction of the earth's crust, and resulting in the production of flexures, foldings, and even inversions, of the strata along lines ranging in approximately parallel and definite directions.

The study of a geological map of this part of the country will probably do more to bring the structure of the strata vividly before the reader's mind than any verbal description. It will be seen that the rocks of this part of Ireland are disposed in long parallel, or sharply wedging bands, ranging nearly east and west,[1]

---

[1] More correctly, a little of south west, and north of east.

and coloured differently according to the formation. The narrower bands are formed of Carboniferous rocks, the broader, of Old Red Sandstone; and while the former are compressed into narrow synclinal, or trough-like folds, forming the valleys and arms of the sea, the latter rise into mountainous tracts, and project as headlands far into the Atlantic, while the strata are bent into great arches, more or less modified by plications somewhat after the fashion of a Moorish arch.

A moment's consideration will suffice to show that the whole of this mountainous tract must once have been covered by Carboniferous rocks, and that where the Old Red Sandstone forms the surface of the ground, it is only because the newer beds have been removed by denudation. And when we further consider that these strata were once horizontally spread in parallel layers over the whole of the south of Ireland,[1] but are now bent into a series of nearly parallel folds and flexures such as I have described, it will be seen that they occupy much less horizontal space than they did originally; that they have in fact been crumpled and bent into these foldings by tangential forces acting in a direction perpendicular to the axes of the flexures;—namely, from north to south, or from south to north.

---

[1] That is, at the close of the Carboniferous period.

It is impossible to conceive, and much less to calculate, the amount of force required to bring about such results. It far exceeds that of gravity, and can only be accounted for on the theory of the secular cooling of the earth's crust. Such a process causes the crust to contract upon itself; and where lines of weakness previously existed, forces the strata into a less horizontal space, and thus to range themselves in parallel folds. These folds can be frequently seen amongst the bare, or slightly clothed, mountains of Kerry; as for instance, in the rugged tract south of Kenmare Bay, and wherever the passes cross the ridges transversely. The massive grits and bands of slate may, in such positions, sometimes be observed to be bent completely over in grand arches, or folded in synclinals, or lastly, rising at high angles into the air.

The district now under consideration includes the loftiest elevations in Ireland. The broken and somewhat serried ridge of Macgillicuddy's Reeks reaches at Carntual[1] a height of 3,414 feet above the sea, and there are several elevations in the same district not much below this. This range stretches westward to the ocean between Dingle Bay on the north, and the long narrow channel of the Kenmare

---

[1] Sometimes spelt 'Carrantuohill,' which Dr. Joyce states to mean 'the inverted reaping-hook.'

River on the south, and rises abruptly from the borders of the beautiful Lakes of Killarney, where we have a combination of natural features rarely to be met with, and rendering this (as all who have visited it in early summer *must* acknowledge) the most delightful spot in the British Isles.[1] The rugged heights of the Reeks, as they slope downwards towards the waters of the lakes, are clothed with forests of timber and underwood, chiefly natural, amongst which the Arbutus is conspicuous, and at the base stretch away the placid lakes, studded with islets, and their banks clothed with verdure. These islets and rocks which rise out of the lakes are often intensely glaciated, rising above the waters, near Muckross Abbey, in long smooth backs of limestone polished and striated.[2] The lakes are situated in the Carboniferous Limestone, but send a long arm southwards into the heart of the mountains of Old Red Sandstone, which terminates in the Black Valley, a gloomy and savage *cul-de-sac*, bounded by steep cliffs stretching along the eastern shoulder of the Reeks.

By the Gap of Dunloe, a narrow gorge which strikes across the ridge into the higher part of the

---

[1] On the high authority of Lord Macaulay. See 'History of England,' vol. ii. p. 307 (edit. 1872).

[2] Amongst which the *Osmunda regalis* grows in fronds of extraordinary size.

FIG. 17.

Section across the Lakes and Mountains of Killarney: to show the structure of the formations and the reversed fold at Killarney. Minor details omitted.

The inner dotted line represents the top of the Old Red Sandstone; the outer, that of the Limestone.

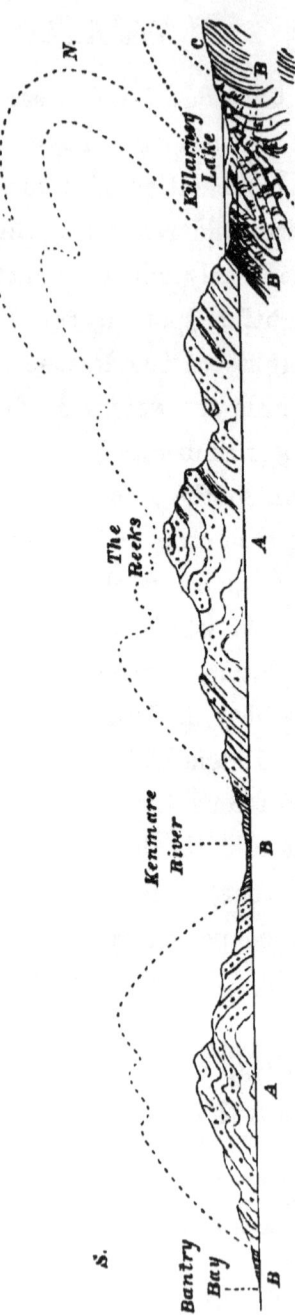

A, Old Red Sandstone.   B, Carboniferous Limestone and Slate.   C, Yoredale Beds (shales).

Black Valley, a fine section of the beds forming the northern flanks of the mountains is obtained. It is here, indeed, that the wonder of the geological observer is excited; for he finds the great beds of green and purple grit of the Old Red formation dipping southwards into the heart of the ridge, and *away from* the Carboniferous Limestone of the plain; and, when he comes to examine the sections of the limestone in the islands and shores of the lake, he finds that this formation dips in a similar direction—that is to say, the newer formation of the plain appears to dip *below* the older formation of the mountains. There is, also, at some distance from the foot of the mountains to the north an inversion of the Yoredale Shales and Carboniferous Limestone. (See fig. 17.)

Here, evidently, is a problem requiring explanation, and one which has occupied the attention of several geologists of eminence, without a perfectly satisfactory solution having been arrived at. The actual junction of the two formations is, I believe, not visible, owing to the presence of a thick deposit of Glacial Drift, and we are obliged to have recourse to theory to account for these phenomena. Two explanations are possible: either the boundary between the Carboniferous Limestone and the Old Red Sandstone is a fault (or fracture of the strata) along which the former formation has been brought down (or displaced relatively to the latter), or else the strata have been folded back upon themselves. This latter is the view adopted with some hesitation by the late Professor Jukes,[1] and is doubtless that which is supported by the greatest amount of evidence. It is probable, however, that at some points along the boundary, which runs in a slightly sinuous line from the southern shore of Dingle Bay eastward, the reversed fold is accompanied by a fracture, or fault.

We infer, therefore, that along such a line of fold or fracture, or both, the beds were thrust up in the form of a grand Moorish arch, the apex of which was between Killarney and Kenmare Bay, where it bent down again to form a narrow trough; again rising into another crenellated arch between Kenmare Bay and Bantry Bay. Between this bay and the southern coast the strata are forced into the arches of Old Red Sandstone forming the headlands of Muntervary, Mizen Head, and Cape Clear, while the intermediate synclinal flexures occupy the valleys which terminate in Dunmanus Bay, and that of Crookhaven or Roaring-water.

The long promontories which jut out into the

---

[1] See Explanation to Sheets 163 and 174 of the Geological Survey, p. 10.

Atlantic along the southern coast are the western extremities of mountain ridges of Old Red Sandstone, which stretch far inland and eastward into the county Cork. Amongst these, the serried ridges of Glengarriff washed by the waters of Bantry Bay are perhaps the most striking; and when seen from the opposite shore against the glowing background of an evening sunset afford studies of shade and colour for the painter, not often surpassed in depth and richness of tone amongst the British Isles.

The outlying mountains of Cork and Waterford, which rise above the limestone plain, often in the form of a dome, or of an oval-shaped ridge, belong to the same system of flexures. They include Slieve Bernagh and the Silver Mine Mountains, elevations which were originally united, but are now intersected by the deep gorge of the Shannon above Limerick; the Maugherslieve, Devil's Bit, and Slieve Bloom Mountains which rise from the central plain, and the grand mass of Galtymore, which is the highest of the group, and reaches an elevation of 3,015 feet. These all have a very similar structure. A central core of Silurian rocks throws off all around beds of Old Red Sandstone and Conglomerate, which in their turn plunge below the Carboniferous Limestone of the plain. In the case of the Galtees, although the general structure is dome-shaped, the form of the mountain is elongated in an east and west direction, corresponding to that of the system of flexures we have been considering.

Geological Age.—And now it is time to consider the question of the Geological Epoch at which the powerful system of terrestrial forces came into play, owing to which the great east and west flexures and foldings of the strata were originated. As the strata which have been influenced by them belong to the Carboniferous group as well as the Old Red Sandstone, it is evident, in the first place, that this epoch was more recent than the Carboniferous itself. But if we were to go further and seek to determine the actual date by reference to the south of Ireland alone, we should be left completely in the dark, because there are no formations in that part of the island newer than the Carboniferous, until we come down to those of the Post-Pliocene period.

On examining, however, the geological structure of the adjoining districts of South Wales and England, and extending our investigations still further into the Continental areas across the English Channel, it becomes evident that the terrestrial disturbances so strikingly represented in the south of Ireland belong to a system which has influenced the Palæozoic strata all over the south of England and Wales, and the borders of France and Belgium to, and even beyond, the valley of the Rhine—a line of several hundred miles. It will be found to correspond with the direction of several of the mountain ranges of the Continent, such as that of the Alps, Pyrenees, and Carpathians; for although these mountain ranges have been affected by disturbances of much more recent date, their earliest movements probably corresponded with those we are now considering.

As far, however, as the British Isles are concerned, we have very precise data for determining the epoch of these movements. In the south-west of England, the New Red Sandstone or Trias is found resting upon the upturned and eroded edges of these older Carboniferous and Devonian rocks. This is the case along the borders of the South Wales and Somersetshire coal-fields, and the shores of the Bristol Channel. The New Red Sandstone formation is therefore evidently more recent than the epoch of the disturbances and erosion of the Carboniferous beds, and we therefore place the date of such disturbances at a period preceding the Triassic.

But we may even go a step further; for in the north of England and Ireland we find the Carboniferous beds arranged in folds and flexures parallel to those of the southern districts. We have, therefore, *the evidence of parallelism of direction* in support of the view that the flexures are contemporaneous. But the flexures of the north were produced prior to the formation of the Permian beds, as I have shown to be the case in Lancashire, and as the late Professor Phillips has shown to be the case in Yorkshire; and we therefore arrive at this general conclusion with regard to these great east and west lines of flexure in the Carboniferous Rocks over the British Islands, that they belong to the period between the Carboniferous and Permian.

This was an age of general disturbance and of great denudation of the previously formed Carboniferous beds, of which thousands of feet in depth were swept off the surface of the land before the deposition of the Permian sandstones and limestones. It was an epoch, therefore, of long duration, but entirely unrepresented by existing records. It is a blank in the stratigraphical series;—a leaf in the history of creation torn out.[1]

[1] In the north of England the Permian beds are found resting directly on the Carboniferous Limestone and 'Yoredale Series,' in which cases all the coal-measures and millstone grit series, amounting from 5,000 to 6,000 feet of strata, had been denuded away. In the north of Ireland at Armagh this formation (the Permian) rests on the Carboniferous Limestone (see p. 46).

CHAPTER V.

NORTH-EASTERN HIGHLANDS.

Mountains of Mourne and Carlingford.—This group of mountains is full of interest as illustrating plutonic and volcanic phenomena of past geologic times, and is remarkable for the number and variety of rocks and minerals which it produces. The Carlingford range is separated from that of Mourne by Carlingford Lough, a deep and picturesque arm of the sea. There is a third line of elevations lying to the north of the Mourne Mountains from which it is separated by a valley of Silurian grits and slates, extending from the Newry Canal towards the north-east, called the Slieve Croob range, attaining an elevation of 1,755 feet, which is composed of granite probably of metamorphic origin, and much more ancient than that of Mourne,[1] to which latter our attention must here be specially directed.

[1] The granite of Slieve Croob is probably of the same age as that of Donegal or Wicklow, and consists of quartz, orthoclase, and black mica.

The mountains of Mourne are composed of a peculiar granite, of which Dr. Haughton has determined the chemical and mineral composition. It consists of orthoclase, albite, quartz and mica, and is full of little cavities which often contain beautifully formed crystals of smoke-quartz, orthoclase, topaz, and more rarely of emerald. It is of intrusive origin, sending dykes and veins into the Lower Silurian rocks with which it comes in contact, and is itself traversed by dykes of basalt, felstone, mica-trap, and porphyry. There are other dykes, however, of similar rocks which are of older date than the granite, and are abruptly terminated at its margin. Thus its age is intermediate between that of the one set of dykes and of the other. These phenomena may be well observed on the hills between Carn Mountain and Slieve Muck, where the contorted Silurian rocks are found capping the granite along the crest of the ridge in a very peculiar manner; one of the 'accidents of denudation.'

The Mourne Mountains consist generally of a series of conical or dome-shaped elevations, of which Slieve Donard, itself forming the culminating summit, is a fine example, rising from the margin of the sea to an elevation of 2,796 feet. This is generally the form assumed by granite in mountainous districts. Still, serried ridges and peaks are not absent from this

The Mourne Mountains. 143

range, examples of which we find in Slieve Bingian (2,449 feet) and Slieve Barnagh (2,394 feet).

The ridge of the Carlingford Mountains presents a peculiarly rocky and serried aspect when seen from the opposite shore of the Lough. It runs parallel to the southern shore of the Lough and reaches in Slieve

FIG. 18.

Diagrammatic section to show how the Syenite breaks through the Pyroxenic Rock.

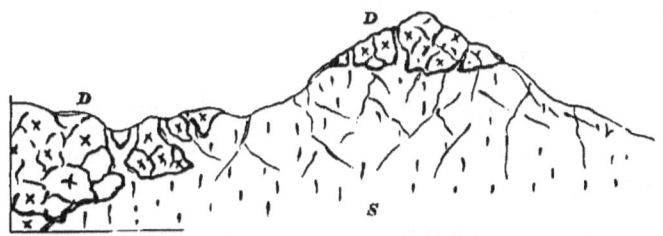

S, Syenite intruded amongst masses of Hypersthene Dolerite, D.

Foy an elevation of nearly 2,000 feet. This ridge is formed of a remarkable hypersthene dolerite, largely crystalline granular, and composed of anorthite, hypersthene, and magnetic iron-ore. The rugged ridge of Barnaveve is capped by a similar rock supported and penetrated by syenite of more recent age, composed of flesh-coloured crystalline felspar with crystals of hornblende and a little quartz. At the western base of this ridge there is a mass of largely crystalline dolerite.

The whole of this group of hills is formed of felspathic and pyroxenic rocks of several forms and varieties graduating into one another, which the officers of the Geological Survey are fully persuaded are the representatives in time of those of the Mourne Mountains on the opposite side of Carlingford Lough. Thus we may classify the two sets as under :—

Pyroxenic group (the more ancient), consisting of micaceous dolerite of Slieve Gullion, diorite of Trumpet Hill, dolerite, and anorthite hypersthene rock of Barnavene and Slieve Foy.

Felspathic group (more recent), consisting of varieties of felstone porphyry, and syenite of the Carlingford Mountains, and representing the granite of Mourne.[1]

These rocks are penetrated by innumerable intrusive sheets and dykes of trap, generally basaltic; some of which, as in the case of the Mourne Mountains, are of Tertiary age, others much older, probably Upper Carboniferous. These intrusive sheets, jutting out along the flanks of Carlingford Mountain (Slieve Foy) and cutting the Silurian beds transversely, impart to it that rocky, terraced aspect which may be noticed even from the northern side of the Lough.

The Carboniferous Limestone laps round the

[1] This view is borne out by the chemical analyses of both rocks as determined by the Rev. Dr. Haughton, of Dublin University.

Mourne and Carlingford Mountains. 145

southern base of Slieve Foy, near Greenore, and at the quarries here the basalt is seen to traverse the limestone, both as intrusive sheets and dykes. It is possible that the sheets may be of Upper Carbonifer-

FIG. 19.

Carlingford Quarry.

Horizontal and inclined dykes of basalt traversing Carboniferous Limestone.

ous, and the dykes of Miocene age belonging to the epoch of volcanic activity in the north of Ireland.[1]

Geological age of the Mourne and Carlingford Mountains.—These mountains are so isolated from strata newer than the Carboniferous that much uncertainty exists regarding the date of their eruption. They may undoubtedly be considered as the roots of volcanic mountains, the trunk and branches of which have been removed by denuding agents;

[1] The great numbers of dykes in this district has led Dr. Haughton to conclude it was a focus of volcanic action.

just as if a mountain like Ætna were to be cut down into a group of hills rising 2,000 or 3,000 feet above the level of the Mediterranean. The intrusive character of the rocks; the association of felspathic and pyroxenic varieties, as in the case of more recent volcanic mountains; the innumerable dykes of trap which radiate from, or traverse, the whole district; all point to this region as having been the seat of great volcanic activity. Nor ought we to omit reference to the remarkable mass of agglomerate, made up (as on the southern flanks of Slieve Gullion) of bombs of granite, which have been torn up from the granitic basis of the hill below, and blown through the throat of an old crater, as conclusive evidence that these rocks in some places were erupted at the surface of the land of the period.

What then, let us inquire, may have been the date of all these volcanic operations? Now, as we find on both shores of Carlingford Bay, the Carboniferous Limestone traversed, as already described, by basaltic dykes which seem to be connected with the larger masses of Slieve Foy: and as on Slieve Foy itself, Dr. Haughton, and more recently the officers of the Geological Survey, found the limestone converted into crystalline marble at its contact with the hypersthene dolerite; it is clear, that the trap rocks are themselves of more recent date than the Lower Carboniferous. But beyond this, we have no direct evidence as to age; nowhere do the Mesozoic strata come in contact with the Mourne or Carlingford rocks. At the same time, on comparing the felspathic and granitic rocks of this district with the Tertiary trachyte porphyry of Antrim and Down, it becomes evident that they are exceedingly dissimilar. There is an appearance of recentness in the latter group which resembles the trachytic lavas of Auvergne, the Rhine, and Hungary, as compared with the former, which is entirely wanting in the Mourne and Carlingford rocks, and I therefore feel satisfied that they are older than the Tertiary period. The actual date seems to lie between the Lower Carboniferous on the one hand, and the close of the Mesozoic period on the other. And recollecting to how large an extent the Permian period was characterised in Europe by volcanic outbursts of which we have examples in the south-west of Scotland, I am disposed to refer those of Mourne and Carlingford to the Permian period itself.

We have now passed in review the different mountain ranges, or groups, of our island, and determined, as far as the evidence admits, the composition of their rock masses, their mode of formation, and date of elevation in the sense already explained. It may be useful to place the conclusions in a tabulated form before we close this chapter, and the following table has been drawn up with this object:—

Mountain Groups of Ireland.

NAME.	ORIGIN AND DIRECTION.	GEOLOGICAL AGE.
(1) North-eastern Highlands of Mourne, Carlingford, and Slieve Gullion. (Most recent.)	Old Volcanic.	Later than Carboniferous; older than the Tertiary. Probably Permian.
2. South-western Highlands of Kerry, Cork, and Waterford.	Mechanical. E. and W.	Between Carboniferous and Permian.
3. Western Highlands of Connemara and West Mayo.	Metamorphic.	Between Lower and Upper Silurian.
4. North-western Highlands of Donegal and Derry.	Metamorphic. NE.	Between Lower and Upper Silurian.
5. South-western Highlands of Wicklow, &c.	Metamorphic and intrusive. NE.	Probably between Lower and Upper Silurian.

CHAPTER VI.

ORIGIN OF THE CENTRAL PLAIN.

NEXT to its mountains, the Great Central Plain is the most striking physical feature in Ireland. As already stated, it stretches right across the island from the coast between Dublin and Dundalk Bays on the east to Galway Bay on the west, between which it seldom exceeds 300 feet in elevation, the average being less. Towards the north it is bounded by ranges of hills generally of no great elevation, stretching from the foot of the Ox Mountains in Mayo, along the southern slopes of the table-land of the Arigna or Slieve-an-Ierin Carboniferous Hills, eastwards to the coast at Dundalk; the margin nearly coinciding with the uprising of the Lower Silurian rocks in the counties of Monaghan and Cavan. Along the west, the boundary is very determinate, as it follows the western shores of Lough Mask and Lough Corrib to Galway Bay, beyond which it assumes the form of a terraced range of low hills in the Burren district of Clare, and thence sends an arm southwards to the estuary of the Shannon at Foynes Harbour. Towards the south, it skirts the group of isolated hills of Old Red Sandstone and Silurian rocks, which, as already described, are the advanced outposts of the mountains of Kerry, Cork and Waterford, and include Slieve Boughta, Slieve Bernagh, Keeper Mountain, Slieve Bloom; and further south, Galtymore, and Slievenaman. Along the east, the boundary skirts the granitic range of Wexford, Wicklow and Dublin, reaching the coast at Dublin Bay along the base of the Three Rock Mountain and Killiney Hill.[1] Throughout the area included in this wide circuit, several isolated hills rise above the general surface of the plain, such as the table-land of the Castlecomer and Killenaule Coal-fields, the Chair of Kildare, formed of Upper Silurian beds, and a few small detached hills towards the north-west. The former of these remains as a monument of the former elevation of the plain.[2] The latter is in some degree due to direct elevation.

Geological Formation of Central Plain.—Throughout the greater portion of its area, the Central Plain

---

[1] The Hill of Howth (Ben Edair), Lambay Island, and the granitic bosses of Rockabill would (if the bed of the sea were laid dry) form a continuation of the eastern margin of the Central Plain.

[2] See fig. 7, p. 42.

is underlaid by the Carboniferous Limestone. Indeed, to such an extent is this the case, that the limits of the Plain, except in the north-western districts, might be taken to coincide with the boundary of the limestone itself, as laid down upon the Geological Maps. As already stated, the strata are nearly horizontal, except near the margins, and in proximity to local disturbances. To the north of Dublin, near Malahide, the beds of the 'Calp series' are greatly disturbed; often violently contorted. Such contortions in limestone are sometimes due to the formation of subterranean caverns and the falling in of the superincumbent beds, and are unconnected with ordinary changes in the earth's crust. The limestone of the Plain, however, is itself only occasionally visible, as the greater portion of the surface is overspread by beds of limestone gravel, or boulder clay, belonging to the Post-Pliocene formation already described,[1] or by shallow lakes and sluggish streams; and the extensive peat-mosses, which extend over large areas, are a still more recent covering, and generally occupy the positions of former shallow lakes.

That the Coal-measures (forming the upper members of the Carboniferous group) once overspread all the area now occupied by Carboniferous lime-

[1] See p. 78.

stone, is a proposition which a geologist is prepared to accept as soon as stated, but to the ordinary reader it will not, perhaps, be so clearly manifest. I may perhaps be allowed to use an illustration drawn from works of human art, in order to make the subject more palpable. If a traveller, visiting the regions of early civilisation in Egypt, Syria, or Babylonia, observes the basement walls of palaces or temples, and the foundations of arches or piers, while numerous blocks of hewn stones are strewn around, it requires no history to convince him that he may be standing on the ruins of a Thebes, a Palmyra, or a Babylon. He knows that where there were the foundations, there also must have been the superstructures. Now the limestone is the basement of the Carboniferous superstructure; and the unvarying sequence of beds, at least within the limits of the British Islands as proved by observation, leads us to this conclusion, that representatives of the upper members of the Carboniferous group were always originally present where the basal beds had been laid down; and that when the former are absent it is only in consequence of their removal by denudation.

In addition to the evidence derived from analogy with other districts, such as those of England and Wales; that, fairly deducible from direct observation in Ireland itself, will prove satisfactory to one who is at all familiar with the process of geological induction. Thus, in several places either on the margin of, or rising from, the Central Plain we find remnants of the Upper Carboniferous strata, which owing to special circumstances (chiefly that they occupy stratigraphical basins or synclinal troughs) have escaped destruction, and like solitary columns amidst the ruins of ancient temples, are monuments of the decay and waste which has reigned around. In this way the little coal-fields of Castlecomer and Killenaule in the south, and those of Arigna, Slieve-an-Ierin, and Tyrone in the north, are interesting as showing what kinds of strata originally overlay the Carboniferous Limestone between their widely separated positions. Now, if we compare the succession of the beds in these places we shall find that they are strictly representative; and the manner in which they are broken off and abruptly truncated along their outcrop, proves that they originally extended to undefined distances beyond their present limits. Let us then compare the succession of the beds from the Carboniferous Limestone upwards in the three districts specified, and we shall see how strictly analogous it is, in each case, though slightly varying in details :—

COMPARATIVE CARBONIFEROUS SECTIONS.[1]

UPPER CARBONIFEROUS SERIES.

Castlecomer (South).	*Arigna (N.W.)*	*Tyrone (N.E.)*[2]
Middle Coal-measures; with beds of coal and freshwater shells (*Anthracosia*).	(Absent through denudation.)	Middle Coal-measures of Coal Island, with seams of coal and bivalve shells (*Anthracosia*).

MIDDLE CARBONIFEROUS SERIES.

Lower, or 'Gannister beds,' with thin coal seams and marine shells (*Goniatites, aviculo-pecten*, &c.).	Lower, or 'Gannister beds,' with thin coals and marine shells (*Goniatites, Posidonia*, &c.).	Lower, or 'Gannister beds,' with a few coal seams and marine shells (*Goniatites, Loxonema, Bellerophon*, &c.).
Flagstones and shales. 'Carlow Flags.'	Millstone grit; flagstones, shales, and coarse grits.	Millstone grit, Coarse quartzose grit, &c. of Drumglass.
'Yoredale beds,' or Upper shale series, with marine shells.	'Yoredale beds,' shale series, with marine shells and ironstone.	'Yoredale beds,' shales with thin bands of limestone and sandstone.

LOWER CARBONIFEROUS SERIES.

Upper ⎫ Middle ⎬ Limestone. Lower ⎭	Upper ⎫ Middle ⎬ Limestone. Lower ⎭	Upper ⎫ Middle ⎬ Limestone. Lower ⎭

From the above table it will be observed that the beds overlying the limestone can be identified with

---

[1] The classification here stated is that proposed by the author in his paper on 'The Upper Limit of the essentially Marine Beds of the Carboniferous System,' &c., Quart. Journ. Geol. Soc. London, No. 132 (1877).

[2] E. T. Hardman, Expl. Mem. of the Geol. Survey (Palæont. notes by W. H. Baily, p. 19).

[3] Ibid., p. 10.

each other in their minute details of stratification; and, as evinced by their fossils, by similarity of conditions of deposition in districts lying in opposite quarters of the Central Plain; so that we have in these two facts a very clear corroboration of the view that these beds were originally continuous.

And now, having thus laid the foundation of our inquiry, and demonstrated, I hope, to the satisfaction of my reader that the Limestone Plain was originally covered throughout its area by Coal-measures, let us see what were the original boundaries of this tract. (See map, fig. 20.)

It is clear, in the first place, that the original Carboniferous tract must have been vastly more extensive than the present. In a southerly direction, it must have included all that mountainous district of Ireland ranging through Kerry, Cork, and Waterford; for, as we have already seen (page 140), these Southern Highlands had no existence *as such* till after the Carboniferous Period had come to a close. The southern margin must therefore have extended *at least* as far as (certainly much farther than) the shores of the ocean. Towards the west and north-west, it was bounded by the more ancient highlands of West Galway, Mayo, and Donegal, and by lands now occupying the bed of the Atlantic. Towards the north-east, it was probably connected with the Carboniferous tracts of Cumberland, Lancashire and North Wales; and in a SE. direction the granitic mountains of Wicklow and Carlow formed the margin. Towards the north-west and south-east, the

FIG. 20.

Sketch-map of Ireland showing original and actual Carboniferous areas.

1. Uncoloured portions supposed not to have been covered by Carboniferous strata.
2. Light-shaded portions show original Carboniferous areas.
3. Darker shaded portions show existing Carboniferous areas.

Carboniferous beds were piled up, layer upon layer, against the flanks of the old crystalline and Palæozoic uplands, until the mountains themselves were in a great degree buried beneath them, so that their present relative elevation may be considered as due to disinterment, the newer strata having since been stript off. The extent of the original and present distribution of the Carboniferous beds is shown in the accompanying little sketch-map (fig. 20).

Plane of Marine Denudation.—Professor Ramsay,[1] and after him more recent writers on physical geology, have shown how the existing mountain tracts of the British Isles, and other parts of the world, resolve themselves into original table-lands, with gently sloping surfaces which were originally planed down by the waves of ancient seas, and upon being elevated into dry land were gradually sculptured into mountains by the excavation of the valleys through the agency of rain, and rivers, and, to a small extent, of glacier ice. The Grampian Mountains of Scotland, when viewed from a commanding position to the southward, and the undulating hills of Central and South Wales, when seen from the southern flanks of Cader Idris, and other points, suggest this view of their origin. It is observed that the summits of the ridges tend to rise to such a level that if a plane surface were stretched across the intervening valleys it would rest upon them, and form a surface slightly sloping in certain definite directions. Such a plane would be independent of the dip of the beds; it would touch them while dipping at all possible

[1] Mem. Geol. Survey of Great Britain, vol. i. p. 297, *et seq.*

angles and in different positions, and is the result of denuding action which has taken place since the beds were tilted or flexured.

Ancient plane surfaces.—Ireland can also furnish examples of the principles here enunciated. The remnants of such an ancient plane surface, or of several such planes formed at successive periods, can be clearly recognised. The well-formed table-land of Slieve Partry, rising from the western shore of Lough Mask to an elevation of about 2,000 feet, and formed of Upper Silurian beds, has evidently been a plane surface before the Carboniferous period, as a little patch of sandstone belonging to this formation is found capping its upper surface; and it is probable that at a later period, to which I shall presently more particularly refer, this plane was re-constituted by a newer denudation.[1]

But the most important 'plane of marine denudation' is that out of which the mountains of the south-west of Ireland were originally developed. As I have more than once stated, these mountains belong to an age newer than the Carboniferous; they include the groups on both sides of the Shannon valley, above Limerick, as also Slieve Bloom, the Galtees, Slievenaman, and the ranges of Cork and Kerry. Now, all these mountains lie in the

[1] See Geol. Survey Map, Sheet 84.

centres (or axes) of domes or arches, which have been thrust upwards above the surrounding tracts by those powerful terrestrial forces I have already described;[1] the intervening spaces being filled up by the more recent Carboniferous strata.

This elevation of some tracts (the arches and domes) would probably be accompanied by a depression of those adjoining; and, as at the close of the Carboniferous period the whole of the surface of the country was but little elevated above the ocean, the result of those movements I have described would be to raise the domes and arches high up into the air, and to slightly submerge the surrounding upper Carboniferous tracts of nearly level strata.

Under such circumstances, the unsubmerged tracts would be exposed to rain and river action; and denudations would ensue, facilitated by the cracks and fissures induced by the bending of the beds into such domes and arches. Meanwhile the waves of the sea would, we may suppose, be operating all along the coast lines, breaking down and reducing to their own level the elevated lands; and if this process of waste by the meteoric as well as by the marine agencies were continued for a sufficient length of time, the whole area would be reduced to one general level; or if the land should be slowly rising or falling, to the

[1] Page 131.

form of a gently sloping plane such as is meant by a 'Plane of marine denudation.'

Now, supposing this plane to be elevated into land, let us see of what it would consist. Evidently of Coal-measures beyond the limits of the domes and arches; and of Carboniferous Limestone, Old Red Sandstone, and in a few instances, of Lower Silurian rocks occupying the central portions or cores; the other beds extending around in parallel zones, and dipping away from the centres and axes of elevation.

Let the reader now turn to a geological map, and he will see that this is exactly the arrangement of the strata among the mountain groups of the south of Ireland on both sides of the valley of the Shannon. To the north of these, the Central Plain would be covered by Upper and Middle Carboniferous beds to a height of 3,000 or 4,000 feet over the existing surface of the limestone, and this nearly level tract would extend towards the north, where the beds had been subjected to somewhat similar, though smaller, disturbances and denudations. And thus from north to south at the close of this long period of terrestrial disturbance and denudation lying between the Carboniferous on the one hand and Permian on the other,[1] the surface of Ireland would present the appearance of a plane, partly formed of coal-measures and partly of

[1] See *ante*, p. 140.

older rocks, with a slight inclination in various directions, along which the streams and rivers would begin to flow whenever the whole tract would be elevated into land.

The highest portion of this plane would be situated (we may suppose) over the region of the Killarney Mountains, whence it would slope away both towards the south and the north, while its lowest depression would lie over the region of the estuary of the Shannon near Foynes. From this it would gradually rise northwards, touching the apices of the Slieve Bernagh (1,746 feet) and Slieve Boughta Mountains, and stretching away towards the table-land of Slieve Partry on the west and of Slieve-an-Ierin and Cuilcagh (2,188 feet) on the north. The slope of the plane from the Cuilcagh table-land towards the Shannon Valley south of Slieve Bernagh *has determined that original course of the Shannon, which it has ever since maintained*; while the gentle slope from the table-land of Slieve Partry inland has probably been the cause why the rivers Shannon and Suck did not elect to flow into the sea at Galway Bay, rather than through the gap between Slieve Bernagh and Slieve Arra as is actually the case;—but to this subject I hope to return.

Towards the north, the original plane probably sloped upwards from the region of Lough Erne in the direction of the Donegal Highlands, culminating over the heights of Slieve Snaght and Errigal; the flanks of these mountains being enveloped in sheets of Carboniferous strata. In a north-easterly direction, the character of the plane is less easily understood than elsewhere, owing to the modifications it has undergone by subsequent depressions and denudations; but towards the south-east, it extended over the culminating ridges of Slieve Bloom, Maugherslieve, and Galtymore, across to the higher elevations of the Wicklow Mountains, being depressed over the intervening Carboniferous tract of Castlecomer and Killenaule. At this time the flanks of the Silurian and granitic elevations of the present day must have been deeply buried beneath masses of Upper Carboniferous strata.

It was, probably, out of such a gently undulating plane, rising over the highest existing elevations, and depressed over the regions of the Carboniferous rocks, that the physical features of Ireland were sculptured. When this plane surface was elevated into land the rivers and streamlets began to collect into those channels which presented the fewest obstacles to their course towards the sea, and these channels once having been selected, were seldom ever after abandoned. Meanwhile, the softer Carboniferous strata began to yield to the influence of the various denuding agents; and terraces, escarpments, and valleys gradually assumed definite form and outline; the models of those which diversify the landscape at the present day.

The Upper and Middle Carboniferous strata, which consist of clays, shales, sandstones, and alternating with beds of coal, would present favourable materials for the agents of waste to act upon: they would be undermined and carried away in a state of mechanical suspension; and when the lower beds, consisting of limestone, were reached, a new process of denudation would come into operation. It is well known that water containing carbonic acid acts chemically on limestone, dissolving it away; and in this manner the waters of rivers, such as the Thames, which traverse calcareous districts, contain various proportions of carbonate of lime. It is easy to see that in such a country as the Central Plain of Ireland, this process would be carried on under very favourable conditions. The numerous streams, fed by abundant rains, and taking up from decaying vegetation a supply of carbonic acid gas, would then, as now, act upon the limestone floor, whether of the Plain, or of the river valleys; and these tracts would be denuded down by the process of chemical solution faster and more uniformly than the hardened Silurian and Old Red rocks of which the mountainous districts are formed.

The mountains were therefore developed, not so directly by the process of upheaval, as by the lowering of the neighbouring tracts.

Duration of the period of Denudation.—If we compare the geological structure of the central and southern parts of England with that of the corresponding tracts of Ireland, we are at once struck with the dissimilarity of their constitution. In the former case, we find the Carboniferous beds passing below successive formations of Upper Palæozoic, Mesozoic, and Cainozoic strata, such as those of the Permian, Triassic, Liassic, Jurassic, Cretaceous and Tertiary periods, which are altogether absent over the centre and south of Ireland. It is only in the north-east that some of these formations are represented, and that sparingly; while others, such as the Jurassic, have no representatives at all.[1] And when we come to put to ourselves the question :—Why there is this difference between the structure of the sister countries? We can only find one answer which is at all satisfactory; namely, that the surface of the Irish area remained in a state of dry land while that of England was being submerged beneath the waters of the sea, over the bed of which nearly all the formations above enumerated were formed.[2] Where-

---

[1] See *ante*, p. 52.

[2] I say 'nearly all,' because in all probability the Triassic strata are of lacustrine origin.

Formation of the Central Plain. 165

ever the sea extends, matter of some sort, either organic or otherwise, is deposited; so that the absence of deposits presupposes terrestrial conditions. It might be suggested, that some of the Mesozoic strata were once deposited over the centre and south of Ireland, and had subsequently been removed; but if so, the probabilities are that some traces of them would have been preserved in the more protected portions of the country. Such not being the case, I prefer the former explanation.

If this be so, how vast was the lapse of time during which this portion of the British Islands was being subjected to the wasting influence of rain, rivers, and other subaerial agents of erosion! During this time the Permian beds, with their varied deposits of sandstone and limestone, stored with marine fossils, were deposited; the great salt lakes of the Triassic period were constructed; successive generations of saurians, molluscs, and other marine forms flourished and passed away during the Liassic stage; great beds of Oolitic limestone, now rising into mountain ridges along both sides of the central axis of the Alps, and composed almost entirely of the shells and skeletons of marine organisms, were laid down over the floor of the Jurassic Sea. Then followed in succession sub-aerial, lacustrine, and estuarine conditions, which at length gave place to fresh submersions under the ocean of the Cretaceous period, during which masses of limestone, many hundreds of feet in thickness, were constructed by the ceaseless industry of lowly organised marine animals, chiefly foraminifera. To these the varied deposits of the Tertiary age were superadded; and over the south of Europe, along a zone extending from the countries bordering the Mediterranean to the frontiers of China, the Nummulite limestone, then greatest limestone formation in the world, was built up mainly of the coiled shells of a special group of foraminifera, the Nummulites. Throughout this inconceivably prolonged lapse of time, our island was more or less unsubmerged, its surface being swept by subaerial waters, and its strata carried little by little into the adjoining ocean, to form perhaps some of the strata which were being piled up over the ocean bed of the British area. At this time Ireland contributed to the future mineral wealth of England; she stript herself to clothe her sister, and to supply materials for protecting from atmospheric waste her vast stores of coal, upon which her greatness and prosperity now so largely depends; this debt ought never to be forgotten.

Owing to the causes above described, can it be wondered at if so small a portion of the original Upper Carboniferous beds, which (as we have seen) once overspread the land, remain to this day; that her originally vast coal-fields are represented by only a few little scraps left here and there, just to tell of former mineral wealth; and that the character of her inhabitants, and their destiny as an agricultural or pastoral people, were fixed altogether independently of social or political considerations?

CHAPTER VII.

ORIGIN OF SOME RIVER VALLEYS.

IRELAND possesses several noble rivers, and her geographical position causes her to receive a plentiful supply of the rains of heaven, so that drought is the last thing she has to fear. Her shores throughout two thirds of their circumference are washed by the waters of the Atlantic, and bordered by mountain ranges which condense the moisture of the prevalent westerly winds; while the rainfall varies from 30 to 60 inches, with an average of about 40 inches per annum.[1] Thus the lakes and streams are well supplied, and form a prominent feature in the physical geography of the country.

Several of these rivers take a course towards the sea which at first sight seems quite unaccountable. Like the Avon at Bristol, the Stour, Medway, and Dart in the south-east of England, they pursue a downward course which appears by no means the easiest or most feasible. We know that water

[1] See Kane's 'Industrial Resources of Ireland,' p. 67 (1844).

will flow by the lowest available channel; and yet such rivers as I have referred to, instead of entering the sea by existing valleys, strike across ridges of comparatively hard or solid formations, through which they have cut ravines; and thus, in what seems an unnecessarily difficult and laborious manner, they effect their escape into the adjoining ocean.

The origin of such river channels has been investigated and satisfactorily explained by Professor Ramsay, Mr. Topley, and other physical geologists as far as regards England,[1] also by the late Professor Jukes in a masterly manner as regards this island.[2] The principles involved are indeed the same for both countries, but require special application in the case of the individual river valleys themselves.

The origin of a river valley cannot be understood simply by examining its course in the light of existing physical features, any more than the institutions of a nation at the present day can be understood unless we examine their rise and development in past times. The early history of both must be searched out if we would desire to arrive at a satisfactory solution regarding their present condition. In the case of the rivers of Ireland, what we have

[1] Ramsay's 'Physical Geography of Great Britain,' edit. 3, p. 166.
[2] Quart. Journal Geol. Soc., vol. xviii. See also Jukes' and Geikie's 'Manual of Geology,' p. 454.

already ascertained regarding the past form of the surface of the country and its geological structure will have prepared us to trace the causes which have led them to assume their existing courses, and we shall find that the early physical condition of the former land surface has in each case predetermined the river channel of the present day.

I shall commence what I have to say under this head with an account of the history of the Shannon Valley, not only because this is the largest river valley in Ireland, but because the account of its origin will throw light upon that of other river channels of smaller size.

CHAPTER VIII.

THE RIVER SHANNON.

THIS great river takes its rise amongst the Carboniferous hills of Leitrim and Fermanagh to the north of Lough Allen, the surface of which is 160 feet above the sea, and on issuing forth from the lough flows sluggishly in a nearly southerly direction over the central plain for a distance of eighty miles, passing through Lough Ree (122 feet) and Lough Derg (108 feet), when it enters the gorge separating Slieve Bernagh from Slieve Arra, and with a rapid fall reaches Limerick, where it becomes a tidal river. Here it turns to the westward and enters the ocean, through a broad and generally deep estuary, about 60 miles in length. The actual head waters of the Shannon are those of the Owenmore, a fine stream with numerous confluents, draining the valley lying between Cuilcagh on the north and Slieve Nakilla on the south, and which flows into the head of Lough Allen. But the traditionary source is a tributary stream which takes its rise in a limestone caldron ('the Shannon Pot'), from which the water rises in a copious fountain. The real source of the water is, however, not at this spot, but in a little lough, situated about a mile from 'the Shannon Pot,' which receives considerable drainage from the ground surrounding it at the base of Tiltibane, but has no visible outlet.[1] The waters from the little lough flow in a subterranean channel till they issue forth at the so-called 'source of the Shannon.'

From the time the river leaves Lough Allen at Drumshambo till it enters Lough Derg, there is nothing remarkable in the physical aspect of its channel. It flows with a slight current, half river, half lake, through the nearly level limestone plain; but here an interesting problem presents itself which we must endeavour to solve. Looking at a geological or orographical map,[2] we find that the central plain here terminates along the base of a group of mountains ranging in a semicircular form from Slieve Boughta on the west to Slieve Bloom on the east. The structure of these mountains has already been explained.[3] They are geological domes, with

---

[1] Mr. S. B. Wilkinson has proved by experiments the truth of this statement, having thrown hay or straw into the little lough, which, on disappearing, has come up in the waters of the Shannon Pot.

[2] Such as that recently published by Mr. Stanford of Charing Cross, London.

[3] Page 118.

cores of Silurian rocks, throwing off the Old Red and Lower Carboniferous beds round their flanks. At Killaloe the Shannon crosses one of these, cutting it into two sections similar to each other in structure, so that without the intersecting gorge of the river they would form a single mountain dome; the channel here consists of Silurian grits and slates followed by Old Red Sandstone—rather hard beds, it must be confessed, for a river to cut through.

Now to the north of these mountains the limestone plain stretches westward to the ocean at Galway Bay; and the question arises, why did the river not turn westward on approaching the base of the mountains and flow out through the plain into this bay?

A little consideration will bring us to the conclusion that if the mountains had been where they are at the time the river began to flow, it would not have selected that course on its way to the ocean. Water must flow down hill, and therefore, when we find a river flowing across a ridge through which it has cut a channel for itself, we must assume that before the channel was cut, the ridge did not exist *as such*, i.e. relatively to the district through which the river flows. We therefore conclude that *when the Shannon selected its channel the ridge of mountains through which it passes was somewhat lower than the plain to the north.*

This brings us back to the epoch of the 'plane of marine denudation,' when (as I have endeavoured to show [1]) the surface of the country had a gentle slope from the region where the Shannon rises southwards. It also, probably, was tilted westwards, in the direction of the Slieve Partry table-land, a tilting which has had much to do with giving a southerly rather than a westerly course to the river when it began to scoop out its channel (see p. 161). At this time the plain to the north was overspread by coal-measures, and was relatively higher than the Slieve Bernagh and Slieve Arra ridge, at that time united. Consequently the river flowed over this ridge as the easiest course towards the ocean, cutting down its channel lower and lower into the Silurian and Old Red strata. Meanwhile, the Carboniferous plain to the northwards was being gradually lowered by denudation, and the strata being softer and less coherent than those of the Slieve Bernagh ridge were more rapidly denuded than the latter. Thus the Slieve Bernagh ridge, with the adjoining domes, were gradually developed, owing to the more rapid waste of the Carboniferous tracts around, till the underlying limestone itself was reached, when the process of mechanical waste was replaced (in great part) by that of chemical solution.

[1] Page 161.

The River Shannon.

Meanwhile, the river, having once selected its channel, never abandoned it; but as the land to the north became lowered, the channel was deepened by the action of the somewhat impetuous current, till ultimately the original ridge was dissevered into two masses, and the existing gorge of the Shannon above the city of Limerick was hollowed out.[1] Such appears to have been the history of Ireland's largest river.[2]

[1] The process of reasoning here adopted is almost identical with that used by Mr. Jukes in his paper already referred to, and having worked the problem out for myself, I was glad to find, on reading this paper, that our views concurred so closely.

[2] According to Sir R. Kane, the Shannon basin has an area of 4,544 square miles, and on estimates supplied by Mr. Mulvany, he calculates that for the 97 feet of fall between Killaloe and Limerick, there is a total of 33,950 horse-power in continuous action, day and night, throughout the year. 'Indus. Res.' p. 75.

CHAPTER IX.

THE BLACKWATER.

THIS is another river which follows an apparently unintelligible course in its progress seaward. It rises in the gently swelling ridge which stretches inland from the Dingle Promontory, formed of Middle Carboniferous strata, and flows in a due easterly direction right across the country, along a valley of Carboniferous Limestone, which forms at Fermoy a trough between rising ground of Old Red Sandstone on either side. This direction it retains as far as about four miles east of Lismore, where it approaches the sea, and is so far just such a channel as might have been anticipated. But at this point the conditions are entirely altered, for instead of continuing to flow eastwards towards the sea at Dungarvan Bay, along the same limestone valley, it turns sharply to the south, and, crossing the ridge of Old Red Sandstone, here called the Drum Hills, it enters the sea at Youghal Harbour.

Professor Jukes supposed that this sudden change

of direction was due to the influx from the northwards of a stream from the Knockmealdown Mountains, which enters the Blackwater near Lismore. If this were so, the change of direction ought to coincide with the influx of the stream from the Knockmealdown Mountains; but, as already stated, the junction is about four miles west of the river bend, and therefore cannot, in my opinion, be properly considered to have originated the deflection of the Blackwater. These mountains are simply part of the anticlinal ridge which bounds the physical trough along which the river has so far flowed. It will be observed that the River Lee makes almost a similar deflection from its normal channel below Cork, and crosses two ridges of Old Red Sandstone at right angles to their direction—one at Passage, and the other at the entrance to Queenstown Harbour at Roches Point. The river might have been supposed to have found an easier access to the sea along the limestone trough which stretches by Castle Martyr to the entrance of Youghal Harbour.[1] The cases of the Lee and Blackwater being similar, a similar explanation is required; and it appears to

[1] Professor Jukes considered that the lateral streams passing through the ravine of Glanmire from the dominant ridge of Old Red Sandstone to the north originally determined the course of the Lee at Cork and Passage when the land was less denuded than at present. *Supra cit.*, p. 395.

me that we must fall back upon the supposition that when these rivers originally began to find their way seaward there was a physical obstruction in each case near where the deflection occurs, which prevented them pursuing their normal eastwardly course. Such an obstruction would arise from the tilting of the surface either by a fault, or by a slight north and south bulging of the strata. In the case of the Blackwater, the upthrow fault which traverses the district in a southerly direction into Whiting Bay, parallel to the channel of the river, would appear to have produced such an obstacle as we are in search of; and the river having once selected its channel, never left it during the gradual lowering of the whole valley since the close of the Carboniferous period. In the case of the River Lee the physical cause of the change of course is not so evident, but has probably been originated by the bending of the beds; and their upheaval along the line of a small fault which passes by Cloyne in a N. and S. direction.[1] A very slight impediment to the eastward flow of the stream, when originally selecting its channel, would deflect it to the southwards, and this deflection having once been made, was ever afterwards maintained.

[1] See Geol. Survey Map, Sheet 196.

CHAPTER X.

THE OWENMORE RIVER, NORTH MAYO.

THIS fine stream, which drains the broad moorlands of Erris, formed of Lower Carboniferous beds, takes a westerly course, and traverses the long ridge of quartzite mountains which range from Nephin on the south-east in a semicircular sweep, till it terminates in the bold headlands which overhang the Atlantic at Benwee. According to the *existing* physical features of this region, the Owenmore would probably run in an exactly opposite direction from the present one. It would take its rise in the quartzite mountains, and flow towards the sea, over the low-lying Carboniferous plain, and enter the sea in Killala Bay. Its actual course, however, can be satisfactorily accounted for by supposing that, when it first began to flow, the Carboniferous plain was at a higher relative level than that part of the quartzite ridge through which its present gorge has been cut. Since then the Carboniferous beds have been lowered by denudation more rapidly than the quartzite but the Owenmore has constantly deepened its channel, so as to prevent the course of the stream from being diverted.

Other rivers in the South-west.—As regards the course of the rivers in the S. W. of Ireland, it will be observed they are directly regulated by the geolo-

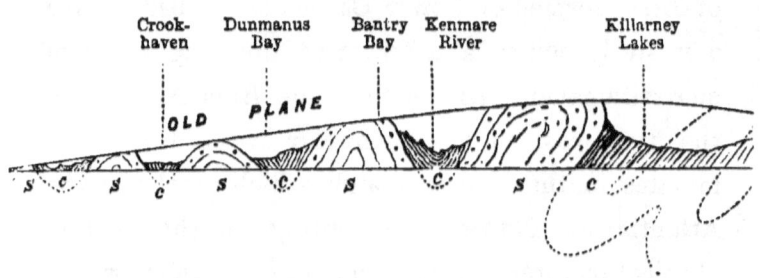

FIG. 21.

Diagram to illustrate the formation of the river-valleys of the South-west of Ireland.

s, Old Red Sandstone forming the mountain arches.
c, Carboniferous beds forming intervening synclinal valleys.
The curved line touching the summits of the arches represents the old 'plane of marine denudation.'

gical structure of the country. They occupy the synclinal troughs of Carboniferous beds which lie alongside the great arches of Old Red Sandstone. The former have ultimately become valleys as well as synclinal troughs, both because of their original position after the flexures were produced, and of their greater destructibility; the softer shales and slates, and the soluble limestones having yielded to the agents of decay more rapidly than the solid grits of the Old Red Sandstone.[1]

During the Glacial period these river valleys were probably less deeply submerged under the waters of the Atlantic than at present. They were probably fresh-water rivers to a greater distance from the interior, and subsequent depression has left them in the form of half river valleys, half arms of the sea. This observation also applies to the estuary of the Shannon, and some of the inlets, like that of Killary Harbour, on the western, as well as to the channels of Cork Harbour and the Blackwater on the southern, coast.

[1] The late Prof. Jukes has recognised the solubility of the limestone in water containing carbonic acid as one of the causes of these valleys. (Quart. Journ. Geol. Soc., vol. xviii. p. 397.)

CHAPTER XI.

OLD DRIED-UP RIVER VALLEYS.

SOME of the most peculiar, and apparently unaccountable, features we meet with in various parts of the country are certain deep clefts crossing ridges, with all the aspect of river valleys, but destitute of streams, either altogether, or of a size corresponding in any measure to the depth and extent of the clefts themselves. Of such a character are the Glen of the Downs in Co. Wicklow; the chasms of 'the Scalp' and Dingle, near Dublin; the cleft which traverses the isolated limestone hill of Keishcorran, south of Sligo; and the Gap of Barnesmore in Co. Donegal. It would be impossible, without unduly lengthening this portion of my subject, to enter into full details regarding the structure, position, and mode of formation of these peculiar features. They are all, as I think, capable of being accounted for on the supposition that they once formed the channels of rivers which flowed through them, when the relative levels of the surrounding country were different from those of the present day. They probably had their origin after the formation of the old 'plane of marine denudation,' when the ground in the directions from which the streams took their rise was at a higher relative level than the existing dried-up channels. Subsequent denudation has altered the relative levels. The ground which once supplied the streams has been lowered, and the channels (from one cause or another) not having been deepened with sufficient rapidity, the streams have forsaken them and have been turned in other directions.

Upon such principles I recently endeavoured to explain the origin of 'the Scalp,' a deep cleft which crosses the granite ridge of Shankill, south of Dublin. For the complete train of reasoning I must refer my reader to the paper itself;[1] but I may here give an outline of the subject.

It is clear that (as I have already shown) the granite ridge of Wicklow and Dublin being more ancient than the Carboniferous period,[2] this ridge was deeply buried beneath newer Carboniferous strata, which were piled up along its sides, layer above layer, till all but the higher elevations were concealed. Upon the elevation of the land, a stream

[1] Read before the Royal Dublin Society during the session 1876–7. 'Scientific Proceedings' for 1877.
[2] See p. 128.

took its rise (probably the infantile Dodder or Shangenagh), which flowed from the northwards over the ground now occupied by 'the Scalp,' and then eastwards out to sea. In process of time by deepening its channel it reached the granite ridge, and ultimately cut out this deep gorge itself. Meanwhile, the land to the north, being formed of Carboniferous strata, was more rapidly wasted and lowered than the more solid granite, and in course of time the relative levels were altered. The land to the north has subsequently been denuded down below the level of the bed of 'the Scalp,' and the stream has in consequence been diverted.[1]

[1] The little rill about a foot or two in width, which now flows along the bed of the Scalp, is evidently not the brook which channelled out this deep and wide chasm.

CHAPTER XII.

ON THE ORIGIN OF THE LAKES.

Lough Neagh and L. Allen.

IRELAND abounds in lakes, both lying amongst her mountains and over the surface of her plains. Those of the former class are analogous to the lochs and tarns of the highlands and uplands of Scotland, England, and Wales; but many of the latter are peculiar to Ireland, both as regards their position and origin. Preeminent amongst all the lakes of Ireland is Lough Neagh, the largest inland sea in the British Isles, and remarkable for its antiquity. For, as we have seen,[1] the beds of clay and sand which occur in such thickness along the southern shores of this lake, and which were deposited under its waters, belong to the Pliocene period, antecedent to the Drift or Glacial times. As the formation of this lake appears to have been of a special character, as well as of older date

---

[1] Page 72.

than that of the others, we shall consider its history first; and I may here premise that all the lakes of Ireland may, with great probability, be classified, as regards their mode of formation, under the three following heads: viz., 1. Lakes of Mechanical origin; 2. Lakes of Glacial origin; and 3. Lakes of Chemical solution. We shall now consider some of them in the order here specified.

1. *Lakes of Mechanical origin.*—Under this head I include lakes which, while they may have been modified in form by other agencies, are primarily due to the faults, or dislocations, of the strata, of which we have two remarkable examples in Lough Neagh and Lough Allen.

The origin of Lough Neagh has been a subject of much speculation, and of some mystery; because, being older than the Glacial Epoch, it cannot be referred to glacial agency, and being situated on deposits other than limestones, it cannot be considered as the result of chemical solution. Its proximity to the old volcanic region of Antrim has naturally led to the inference that it was in some way connected with local sinking of the surface through volcanic agency. It was not, however, till the geological structure of the adjoining districts of Tyrone on one side, and Antrim on the other, had been accurately laid down on the maps of the

Lake Basins.

Geological Survey,[1] that a key to the history of its origin was found, and Mr. E. T. Hardman, one of the officers of the Survey, has very ably applied the results of his examination of the district surrounding the lough to the determination of its mode of formation,[2] and has come to conclusions which (as he informs me) have received the assent of Professor Ramsay, the author of the theory of the glacial origin of lakes.

This great sheet of water washes portions of the counties of Derry, Antrim, Down, Armagh and Tyrone. Its northern portion is bounded by the Miocene basalts of Antrim; its southern partly by alluvial tracts, partly by masses of drift resting on Pliocene clays, which in turn overlie the Triassic or Carboniferous strata. Its length from north to south is 15 miles, and its breadth 12, giving an area of nearly 150 square miles. The general depth is only from 20 to 40 feet, gradually increasing towards the northern shore; and the surface level is 48 feet above that of the sea.

Mr. Hardman shows that along the southern shores, the Pliocene clays, originally deposited under

---

[1] Sheets 27, 28, 35 and 36.
[2] 'On the Age and Mode of Formation of Lough Neagh.' Journ. Roy. Geol. Soc. Ireland, vol. iv. p. 170. Also Explan. Mem. to Sheet 36 of the Geological Survey Maps, p. 72, *et seq.* (1877.)

the waters of the lake, rise to a level of 120 feet above the sea, or 72 feet above the existing surface of the lake, showing how very much greater the area of the lake must have originally been in this direction.

During the progress of the survey, it was found that the strata on both sides of the lake are traversed by several large faults, ranging in ENE. directions; one of these ranges through the basaltic plateau of Antrim by Temple-Patrick, where the vertical displacement is about 500 feet, the downthrow being on the south side. These faults are later than the basaltic sheets of Miocene age which they displace, and of older date than the Pliocene clays, which are not affected by them; the ground having been smoothed down, and the inequalities caused by the dislocation of the beds having been worn away by denuding agencies before the clays were deposited. It was to the depression of the surface through the agency of these faults that, according to Mr. Hardman, the formation of the lake is due; and at this time a considerable river (the progenitor of the Upper Bann and Blackwater streams) entered the lough from the southward, depositing the clays over its former bed where now we find them. During the movements of the surface which subsequently ensued, the southern tract has received a slight tilting upwards, which has laid dry the old bed of the lake in this direction. This lake, therefore, forms an illustration of a basin formed by the mechanical action of faults in the strata, assisted by the action of running water. The production of such faults may not be unconnected with the previous volcanic disturbances to which the region had been subjected, though the igneous outbursts had themselves long previously ceased; and the faults appear to correspond with the lines of more ancient fractures.

Lough Allen.—When describing the source of the Shannon I had occasion to refer to this fine sheet of water, from which the river issues forth at Drumshambo. It forms, in fact, a great head reservoir for the Shannon, collecting the numerous streams draining the high moorlands and wide valleys of Carboniferous shales and grits, which rise to the northwards and on either side of the lough. If the waters of the lake were drawn off, we should have a wide valley, into which several others converge, the banks formed of Yoredale shales and the floor of limestone. It would thus be a great river valley continuous with the Shannon, and this (as I suppose) was its original condition. But a barrier having been thrown across, or rather raised up, at its lower extremity, it has been converted into a lake, and thus resembles one of those great reservoirs which have been constructed amongst the Carboniferous valleys of Lancashire and

Yorkshire for the supply of the neighbouring towns and cities. If so, the question arises, how was this barrier constructed, and of what is it formed?

Now, on consulting the Geological Survey maps of the district,[1] it will be seen that a large fault passes in a direction about E. 20° N. across the lower end of the lake, having a 'down-throw' (or vertical depression) on the north side, so that the Carboniferous limestone is brought down on that side of the fault against the Old Red Sandstone and Silurian beds on the south side. This fault has in fact raised up the barrier which was required in order to convert the river valley into a lake basin. The effect is the same as if a solid wall of slate, sandstone, and limestone had been built up across the entrance to the valley of the Arigna River. Through such a barrier the Shannon has cut its channel. The barrier itself has, indeed, been much lowered and worn down by denudation, and possibly by glacial erosion; but its superior elevation relatively to the bed of Lough Allen has been to some extent retained, so that the waters are pent up and the lake still remains.

What may be the age of this lake, or at what period the fault was produced, we have no means of knowing. The fault is more recent than any of the strata of the district, and as its direction is very similar to

[1] Sheets 66 and 67.

that of the Post-Miocene faults of Antrim (to the existence of which Lough Neagh seems to have originally owed its origin),[1] it is not at all improbable that it was produced at the same epoch of terrestrial movements with them; namely, between the Miocene and Pliocene periods. It is, in fact, very nearly continuous with a great line of displacement which ranges by Swanlinbar, and which, after crossing the Lough Erne valley by Knockninny, ranges by Lisnaskea and Slieve Beagh towards Lough Neagh by Coal-Island. The actual fracture may not be continuous throughout, but the separate fractures of the strata probably belong to the same system of disturbances.

[1] See *ante*, p. 188.

CHAPTER XIII.

LAKES OF GLACIAL ORIGIN.

These lakes are to be found chiefly amongst the glens of our mountains, or in front of the valleys, where glacier ice has debouched on the plains; they occur in immense numbers and of all sizes amongst the Donegal, Mayo, Galway, and Kerry Highlands. They are also found in certain districts which have been only subjected to ice erosion during the period of general glaciation, as for instance in the hilly districts of Westmeath, Cavan, and Monaghan. But this glacial origin of lakes is of two kinds:— first, as in the cases where hollows have been actually scooped out in the solid rocks by the passage of ice over the surface; and, secondly, where moraine matter, or boulder clay, has been heaped up across a valley or hollow so as to form an embankment for the streams which enter the depression from above.

Perhaps there is no district in Ireland where rock basins and moraine-dammed lochs are so numerous as in that which lies at the foot of the Twelve Bins of Connemara. A glance at the Ordnance, or larger geological maps will illustrate this better than any description. Some tracts in this country, such as that lying to the south of Clifden and the tracts bordering Kilkerran Bay, are a perfect network of loughlets, ice-worn bosses of rock, and drift mounds. The coast itself consists of irregular and deeply indented mountains and inlets, studded by sunken rocks and islands, such as would arise from the partial submersion of the neighbouring lands. Now we know from observation that the whole of this district is intensely glaciated, and has been subjected to ice-erosion during two periods;—that of the general glaciation and that of the more recent local glaciation.[1] The Geological Survey maps of this district, prepared by Mr. Kinahan, afford abundant evidence of this by the numerous glacial marks which have been inserted on them over the districts referred to; and it is impossible by any other theory than that of glacial agency[2] to account for the rock-basins and chains of loughlets by which this country is diversified, and which, when viewed from some of the neighbouring elevations, produce on the mind an impression of a district over which land and water are struggling for the mastery, but where the water is gradually getting the better of the land.

[1] See p. 242.   [2] As proposed originally by Prof. Ramsay.

Of a similar character is the tract stretching from the base of Slieve Snaght in Donegal to the shores of the Atlantic, which has been referred to by Mr. J. F. Campbell as one over which former ice-action is especially apparent. The granitic moorlands are studded with loughs and loughlets, while the deeply indented coast is bordered by innumerable islands, the largest of which, Aran Island, faces the continuous swell of the Atlantic.

The mountains of Kerry also contain numerous lakes and mountain tarns, the origin of which can only be attributed to glacial agency under one form or another. The narrow rock-basin of the Upper Lake of Killarney, which fills the bed of the deep gorge of the Black Valley, is itself in the line of an ancient glacier which descended from the base of the Reeks, and debouched on the limestone plain. The larger of these lakes is situated on the Limestone formation—and therefore owes its origin (or at least its great extension) to another cause presently to be referred to; but the glaciated surfaces of the rocks along the shore, and the position of the lower of the lakes in front of the entrance to the Black Valley, leave no room for doubt that they have been deepened by the grinding action of glacial ice. The mountainous promontories lying between Dingle, Kenmare, and Bantry Bays enclose amidst

Lake Basins.

their glens and hollows many loughs and tarns of similar origin. I cannot, however, do more than give an example; but, *ex uno disce omnes*—no one can doubt that, in a region which has been the seat of glacial action on a grand scale, such sheets of water are attributable to this cause directly or otherwise.

The example now to be described occurs amongst the mountains of Kerry overlooking Kenmare River. A wide valley descends from the mountains above Sneem and enters the northern shore of the Kenmare estuary, along which the evidences of glacial action are frequent and striking, and have received special notice from the pen of the Rev. M. Close,[1] and Mr. Wynne of the Geological Survey.[2] The stream of ice which moved down the Kenmare River Valley received a tributary from the Sneem basin, as indicated by the rounded and scored rocks at, and above, Sneem village; but (as Mr. Close observes) some of the ice of that valley came across the ridge at the head of it from the upper part of Glencar, as the ice grovings were observed by Mr. Wynne to cross the ridge transversely in the vicinity of a little mountain tarn called Lough Coomanassig. That little lough is 1,500 feet above the sea, and lies in a rock-basin the

---

[1] *Supra cit.*, p. 226.
[2] 'Explanation,' Sheet 182 of the Geol. Survey maps.

lower side, or lip, of which is greatly rounded away. It appears to have been scooped out by a small local glacier, which, at 500 feet lower down, threw a terminal moraine across the valley, and thus originated a second mountain tarn called 'the Eagle's Lough.' Here, then, we have in proximity examples of loughs owing their origin to glacial agency under both its forms.

In considering the origin of some of the large lakes of the Central Plain, and neighbouring districts, situated in part, or altogether, in Carboniferous limestone hollows, though their origin is mainly due to other causes (as we shall see) than that of glacial erosion,[1] it is impossible to doubt that their form and extent is in some degree attributable to this cause. These lakes are in fact true rock-basins. At their outlets the solid rock rises into the river bed, and is at a higher level than the deeper portions of the lake bottom itself. Amongst these may be named Lough Erne, Lough Corrib, Lough Mask, Lough Gill, and Lough Derg.

Amongst the lakes of glacial origin belonging to the Wicklow range, may be specially named those of L. Bray and L. Nahanagan. This latter is a remarkable example of a moraine-formed mountain tarn; that it is also a rock-basin there can be little

[1] See further on, p. 198.

doubt, but not knowing its depth, I cannot state this with confidence. In August 1877, I visited this Lough with Professor Ramsay. It lies at the head of the Vale of Glendasan in Co. Wicklow at an elevation of about 1,000 feet above the sea-level. It is bounded through about half its western margin by a cirque of granite cliffs, rising several hundred feet from the water edge, and along its opposite side by a well-formed moraine of granite *débris*, which has recently been laid open to a depth of fifty feet below the outfall. All the way up the glen the granite rocks are remarkably glaciated by ice which has moved in an ESE. direction, the glaciation extending to an elevation of 1,800 or 2,000 feet above the sea. These ice-worn bosses are specially remarkable near the lead mines.

After careful examination I have also come to the conclusion that the Round Tower and churches at Glendalough are built on a moraine, which has been thrown across the Glendalough Valley by the glacier that descended the vale of Glendasan. This moraine had originally pent up the waters of the lakes, and against its northern flank the old terrace of gravel (described in another page) has been deposited. Afterwards the river cut down its channel and lowered the level of the waters. In this view Professor Ramsay concurs.[1]

[1] We examined this place together, August 16, 1877.

CHAPTER XIV.

LAKES DUE TO CHEMICAL SOLUTION.

ANYONE who examines the map of Ireland attentively cannot fail to be struck by the great number of lakes which are distributed over the Central Plain and the adjoining districts to the northwards lying inwards from the Bays of Sligo and Donegal; and if it be a geological map he will also notice that these lake districts are chiefly situated on tracts occupied by the Carboniferous Limestone formation, or along its margin.

Some of these lakes are of large size, only exceeded in extent by Lough Neagh, the origin of which is (as we have seen) wholly exceptional. In the north-west beyond the borders of the Central Plain are Lough Erne (Upper and Lower), Loughs Melvin, Gill, Arrow, Garra, and Conn. Some of these, however, invade districts partly composed of other than calcareous rocks, a fact for which we have to account. In the Central Plain itself, there are the Longford and Westmeath Lakes situated wholly upon limestone, and celebrated for their large trout; amongst these may be specially mentioned Loughs Sheelin, Lene, Derravaragh, Iron, Owel, Ennell and Lough Ree, the largest of all, through which the Shannon finds its way from north to south. Bordering the Central Plain to the westward, are the great chain of lakes, Loughs Carra, Mask, and Corrib; the first of which is entirely in limestone, the other two having portions of their western shores formed of other and older rocks. Then southwards, there is the long sheet of water through which flows the Shannon for a distance of twenty miles; lying also in the limestone, with the exception of its southern extremity which is channelled out in the Silurian slate, forming there the wide bed of the river. All these last-named great sheets are true rock-basins. In the limestone tract lying to the west of the Slieve Bernagh Mountains, and extending to the River Fergus, there are several loughs of smaller extent; and, lastly, at the extreme south of the limestone district, lying at the foot of the Reeks, are the lakes of Killarney.

In examining the form of these lakes—the manner in which they widen out in some places, in others become contracted—it will generally be found that they spread themselves out over the ground formed of limestone, and contract where non-calcareous rocks form the bed and margin of the lake. Lough Derg offers an illustration of this, as it will be observed how it spreads out in both directions over the limestone plain, before entering the gorge above Killaloe, where the banks of the lough (here merging into the river) are of slate.[1] L. Erne offers another example as it spreads out considerably in the direction of Kesh, where its bed is limestone from side to side. In the case of the Upper Lake the waters have expanded right and left, chiefly over limestone ground. The district we are now in, extending from Enniskillen to Cavan, generally called 'the valley of the Upper Erne,' presents, when seen from an elevation, a wonderful labyrinth of winding lakes and streams, separated by mounds and banks of drift, amongst which the river sluggishly winds, or sends out arms and loops, which sometimes widen out into loughlets. Part of this surface configuration is doubtless due to the irregular distribution of Boulder clay and gravel, accumulated by the great ice-stream which moved northwards along the valley of the Erne from the central snowfield; but another cause is the solubility of

---

[1] See p. 172. Lough Gill is a true rock-basin lying in a limestone district, and, like many others similarly placed, is due to chemical solution preceded by glacial erosion. Its depth is 90 feet.

limestone under acidulated water, which I now proceed to explain more fully.

When treating of the formation of the river valleys of the South of Ireland, I mentioned as a cause of the formation of such valleys the solubility of limestone in presence of water containing carbonic acid gas. That the existence of the lakes in the district we are now treating of is mainly due to this cause there can be no doubt; the large numbers that are distributed over the plains, or along the margin of the Central Plain, where (as in the case of Lower Lough Erne, Lough Mask, and Lough Corrib) the limestone abuts against the older non-calcareous rocks; and the phenomena of widening to which I have just referred, all concur in leading to this conclusion. These lakes are, strictly speaking, irregular hollows dissolved out of the limestone floor, and filled with water. Their sides are often most irregular in form, and when laid down on a map strongly suggest the idea of encroachment of the water on the shores by a process of melting. The manner in which the limestone itself, when exposed along the margins of the lakes, or on the surface of the ground, dissolves away under meteoric influences is sometimes beautifully illustrated on a small scale, as in the case of the eastern shores and borders of Lough Mask. The rain-water percolating along the vertical joints and horizontal beds dissolves the surfaces into fantastic forms—sometimes like tables, anvils, and little pillars, gradually widening the openings till the whole mass ultimately disappears. In this way the rivers and streams which drain the Central Plain carry away enormous quantities of carbonate of lime in solution, and by this process not only are the lakes being constantly enlarged, but the general surface of the limestone plain is being slowly lowered.

The flatness of the country and the sluggishness of the waters enable them to act with greater effect in dissolving the calcareous rocks. The limestone itself is often penetrated by underground rivers, which create hollows arched over by the rock; but when these give way—as sometimes is the case—a chasm is created, and the commencement of a lake may be the result. Thus the waters of Lough Mask flow into the head of Lough Corrib in one or more underground rivers channelled through the beds of limestone, and issue forth at Cong under the limestone cliff in several magnificent fountains.

CHAPTER XV.

UNDERGROUND RIVERS.

IN connection with this subject I may refer to the underground river channels in the limestone district of Sligo and Fermanagh. A complete description of them would form an interesting essay in itself. In this district the Carboniferous beds, as previously stated, rise into terraces and tabulated moorlands. The higher portions of these hills are often formed of Millstone Grit and Yoredale beds, which collect the rain and send down streams on to the limestone platforms below. On reaching the limestone, however, the water-tight bed fails, and down sink the streams into the honey-combed calcareous rock, through which they penetrate, till they issue forth in copious fountains. One of the most remarkable examples is the fountain called 'the Marble Arch' in Florence Court Park, the origin and structure of which I have attempted to illustrate in the geological section (fig. 22). In other cases, the rivers have failed, owing to changes in the configuration of the

FIG. 22.

Section through Cuilcagh and 'the Marble Arch,' Florence Court.

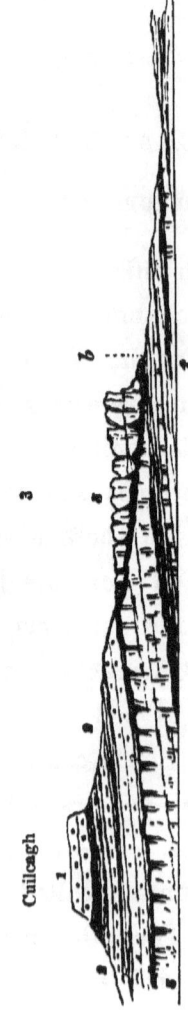

a, Position where the streams disappear and descend into the Upper Limestone (3).
b, Position of Marble Arch and fountains, where the Upper Limestone (3) rests upon the Middle Limestone, or Calp shales, &c.

1. Millstone Grit. 2. Yoredale Beds. 3. Upper Limestone. 4. Middle or 'Calp' Beds. The dark band shows the position of the underground channel.

country, in which case we have ramifying caverns, such as those with which Belmore Mountain near Enniskillen is perforated.

From what I have said it will be seen that there is a close analogy between underground river channels and lakes in limestone districts. The former are dissolved out by water beneath the surface, the latter by water at the surface; and the one may ultimately be transformed into the other. Now, although underground river channels and dry caverns are as plentiful in England as they are in Ireland, as for example in Derbyshire and Yorkshire, yet in the former country there are no such lakes occupying basins in limestone, either of Carboniferous, Jurassic, or Cretaceous age as those here described. Nor is it difficult to assign a reason for this contrast. In England, the limestones nowhere form plains nearly level and slightly elevated above the sea. The Carboniferous Limestone is there 'the Mountain Limestone,' always rising into hills, ridges, or scarped table-lands; so with the Oolite Limestones; so with the Chalk. Consequently, the rain which falls upon their surfaces either sinks down through crevices or fissures, and flows off through underground channels into rivers, or runs directly off the surface into rivulets and streams, which drain into the sea. Thus the rain has no time or opportunity to settle in hollows, and gradually corrode the rock into caldrons, meres, or lakes. With a steep fall in one direction or another, it is drained quickly off;—which, as I have shown, is a state of things very different from that to be found in the Central Plain of Ireland.[1]

I have said above, that we must account for the fact that, as in the cases of Loughs Corrib and Derg, the waters of the limestone lakes sometimes invade tracts occupied by other kinds of strata. In some instances this is probably due to mechanical wear and tear of the banks, after the limestone has all been dissolved and the marginal strata reached by the action of the waves, or currents of the lake itself—a kind of *lacustrine denudation*. This phrase may be a new one, but it is nevertheless true. Let anyone who doubts it sail down Lough Corrib in the little steamer that plies between Cong and Galway, on a stormy day, and he will find that there can be waves, aye, and 'a high sea,' on these big inland lakes; and if waves, then abrasion along the shores, where these waves break on the rocks, and thus an extension of water-territory at the expense of the land. In this way we may often account for shallow indentations of the non-calcareous shores by the waters of these

[1] The Cheshire meres bear an outward resemblance to some of the central Irish lakes, but are entirely different as regards their physical position, being situated on the New Red Marl of the Triassic formation.

limestone lakes; but where these indentations extend downwards into deep water such an explanation will not suffice, and we fall back upon 'glacial erosion,' during the period of the great ice-sheet, as the cause in such cases.

CHAPTER XVI.

BAYS AND INLETS IN LIMESTONE DISTRICTS.

CLOSELY connected with the subject of the chemical solution of limestone is the formation of many of the bays and inlets around the coast of Ireland, which have been formed where the Carboniferous Limestone borders the sea. In many cases (as those of Kerry, in the S. W. of the country [1]) these bays and channels are to a great extent submerged river valleys; in other districts, this is not so clearly the case; but on the other hand a clear connection may be traced between the range and extent of the limestone and the indentation of the coast, quite independent of the river valleys which may open out upon it at that place. Thus along the western coast there are the Bays of Donegal, Sligo (with several deep indentations running into the limestone coast), Killala, Clew Bay, Galway, Tralee, and Dingle; along the south, Dungarvan; and along the east coast, Wexford, Dublin, and Dundalk Bays. The

[1] See p. 181.

connection between the limestone of the coast and the extent and position of these bays will be plain to anyone who studies the formation of the coast by the aid of a geological map. He will find that where the bays are most deeply indented, there the limestone prevails; while on the other hand the margins and headlands are formed of older non-calcareous rocks.

It is impossible to doubt, therefore, that the bays along the coast have in such cases been formed by a process similar to that which has hollowed out the basins of so many lakes over the Central Plain; namely, that of chemical solution. The waters of the sea, as we know by the researches of Bischof and other chemists, contain sensible quantities of carbonic acid, derived from the decay of animal or vegetable matters; and these, acting on the soluble calcareous rocks of the coast, gradually dissolve them, and in process of time produce the indentations or bays, which are so useful as sheltered sites for cities and towns, and as natural harbours of refuge for ships.

But whatever the extent and numbers of the existing lakes of the central districts, there is reason to believe they were once greater. Most of the great peat-mosses which overspread so large an extent of the country occupy the positions of former lakes. They are generally underlaid by white calcareous marl, formed of decomposed fresh-water shells, which is sufficient proof that they were lake beds. Since that time, however, the vegetation of the borders has gradually encroached on the domain of the waters, and has often entirely, at other times partially, replaced the waters of the original lakes.

We have now passed in review the various ways in which the valleys, lakes, and bays of Ireland may have been produced, and we now proceed to examine the effects of that powerful glacial action which has so strikingly modified the form and features of the country, over the plains as well as the hills and mountains.

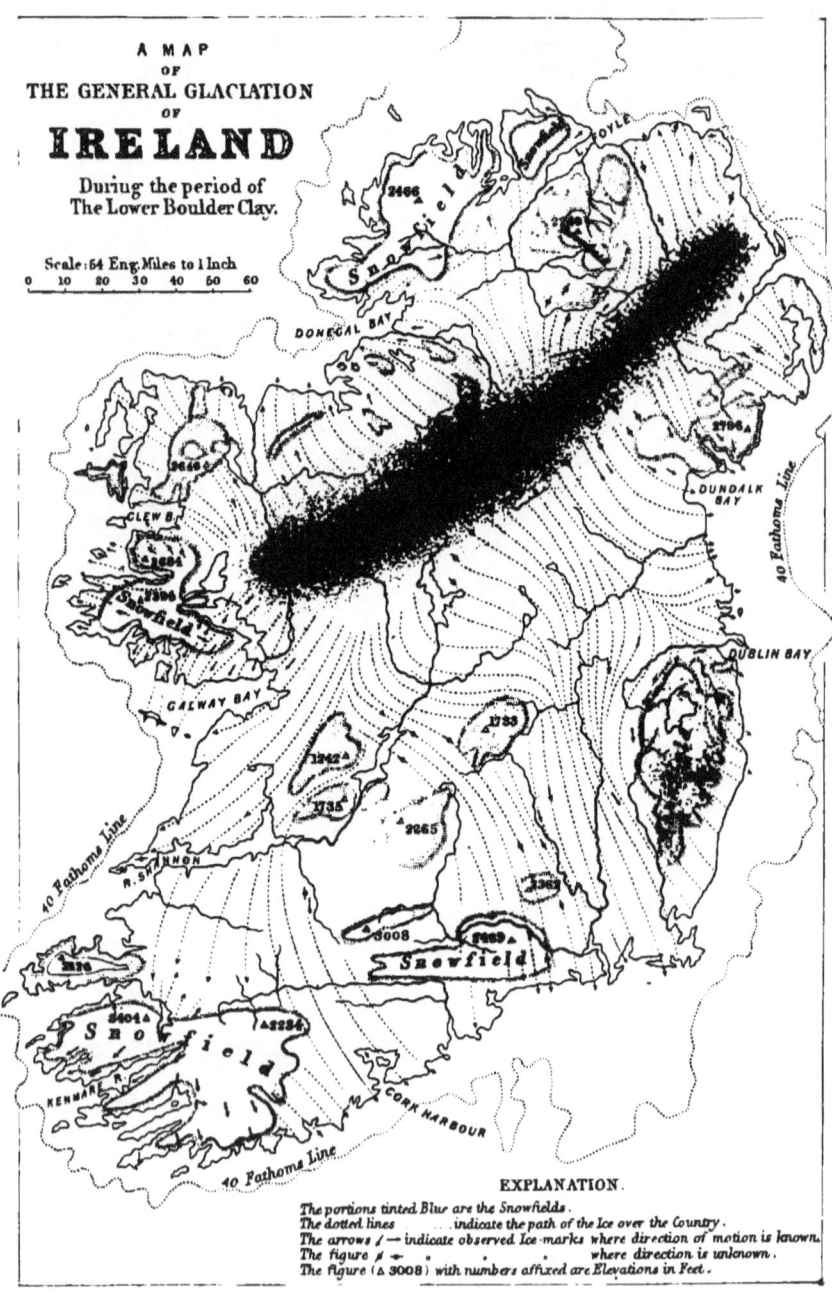

PART III.

THE GLACIATION OF IRELAND.

CHAPTER I.

GLACIAL EVIDENCES.

THAT the surface of the solid rocks of Ireland are extensively glaciated—in other words, bear the marks of glacier-erosion, has for some time past been generally recognised.[1] But the determination of the broad principles which governed the movements of an ice-sheet over the surface of the country is mainly the result of investigations by the Rev. Maxwell Close, who in his admirable papers on the subject of the 'Glaciation of Ireland' has brought together the observations made by himself and the officers of the Geological Survey, and has enabled us to obtain very clear views regarding the position, both of the reservoirs of snow which fed the streams of

[1] See Lyell's 'Antiquity of Man,' 4th edit. Geikie's 'Great Ice Age,' 2nd edit. p. 394.

ice, and of the directions in which those streams flowed during the period of general glaciation.[1]

In dealing with this subject, it will be necessary for the reader to bear in mind the distinction between *general glaciation* and *local glaciation*; the former having predominated over large tracts, not only of the mountains but of the plains, and representing the physical condition of the ice during the period of the Lower Boulder Clay or Till; the latter being confined to local sources of glacial streams descending towards the plains from the higher elevations of the isolated mountain groups, and belonging to a later period. This being so, it will be necessary to treat the subject under its two divisions of *general* and *local glaciation*, both of which have left their traces on the rock surfaces in the form of polished and moulded bosses, scars, delicate chiselings, or deep groovings often intersecting each other at various angles, and requiring care and some amount of general knowledge of their causes, in order to distinguish to which period of glaciation they belong.

When describing the Lower Boulder Clay,[2] I noticed the phenomena generally observable at its contact with the solid rock on which it is found to

---

[1] See References to Authors, p. 257. Mr. G. H. Kinahan and Mr. J. F. Campbell have also contributed much to our knowledge of local details of ice-action.

[2] See p. 79.

rest. I stated that the surface of the rock is often polished, and scored by fine lines and parallel groovings, marking the direction in which the ice-sheet which originated the Boulder Clay had moved. The mapping of these lines and grooves all over the country has enabled us to determine the general nature of this ice-movement; that it belongs to one grand dominant system of central pressure, which has caused the ice to move outwards in all directions from a central snow-field towards the existing coast-lines, except where impeded or deflected by local mountain barriers. The path of the ice-sheet from its source is represented in the accompanying little map, which the reader will do well to consult from time to time as he reads these pages (map, p. 224).

How such ice-marks were produced is now well understood, and need not here be described further than in a few lines. We take our lesson from the existing centres of glacial action, such as the Alps, Norway, and Greenland; and we there find that the sides and floors of valleys or plains, occupied, or liable to be invaded, by moving ice, present precisely similar appearances to those we find amongst our own mountains, and on the neighbouring plains from which the ice has now altogether disappeared. We cannot doubt, therefore, that the phenomena we observe at home are the result of causes similar to those which are in operation abroad; and the question arises, how does moving ice produce these surface groovings and striations? In answering this question we have to observe, that glacier ice, when descending the valleys, receives on its surface blocks and fragments of rock from the cliffs which rise on either hand; some of which fall through and become imbedded in the bottom of the glacier. Other fragments, angular or otherwise, are caught up by the ice as it moves along; while the glacier ice holds small particles of sand and clay, which give the muddy character to the water issuing forth from a glacier valley, such as that of the Rhone before it enters the Lake of Geneva.

Now, all this sand, mud, stones, and boulders, firmly imbedded in the moving ice, and acting on the floor of solid rock over which it slides, causes it to become a great polishing and grinding engine; and where it has a solid, unyielding surface of rock to act upon—such as granite, basalt, quartzite, or limestone, it leaves its deepest and most lasting traces; sometimes wearing down their irregularities, smoothing and polishing the surface; and lastly, engraving upon it strong lines and furrows to denote the direction in which the grinding engine was moving when engaged in its work.

In such a way has the ice-trail been left where

Glacial Evidences.

the rock was solid, and protected from weathering by a coating of boulder clay however slight. But when the underlying rocks happen to have been of a soft or fissile character, the effects have been different. In such cases the gliding of the ice-sheet over their surfaces has usually resulted in breaking up, or else bending the strata over in the direction of the

FIG. 23.

Strata of Silurian grit and slate bent over in a direction opposite to the dip, as is supposed, by the movement of an ice-sheet. Rathdrum Station, Co. Wicklow. Height 30 feet, length 90 feet.

motion. For example, the district lying to the east of the Wicklow Mountains consists of slaty rocks in some places, and in others of quartzites. In the latter cases, as at Carrick Mountain and Bray Head, the surface of the rock is often polished so as to glisten in the sun's rays, and preserves the faint lines of glacial erosion with sufficient distinctness for observation. But in the case of the slate districts, the slates are generally broken up, crumpled, or bent over, as may be seen at Rathdrum Station in Co. Wicklow (see fig. 23) and many other places.

Ice-worn Rocks.—Another mode of determining the existence and direction of former ice-movements is by noting the form of the bosses of rock, which are often to be observed rising above the sward of the plains, or projecting out of the heather or peat of the mountains (figs. 24 and 26). Sometimes these sur-

Fig. 24.

Ice-worn bosses (*roches moutonnées*) of basalt on the summit of Ben More or Fair Head, overlooking Ballycastle Bay at an elevation of 630 feet.

faces still retain ice-marks, but oftener not so; and in this case the shape of the boss enables us to infer, not only the rectilinear direction of the ice-movement, but also that point of the compass towards which it moved, and from which it came. Such bosses, called by the French and Swiss geologists '*roches moutonnées*,' often have a 'crag and tail' form; that is, one end terminating abruptly, the other pointed and gently sloping downwards to the ground. It is evident that the blunt or 'crag' end has been subjected to less attrition than the 'tail' end, and hence we conclude that the crag-end is turned in the direction towards which the ice had moved. Beautiful examples of such *roches moutonnées* may be observed at the Killarney Lakes, at Glendalough in Connemara, at Pontoon,[1] a narrow channel between Lough Conn and Lough Cullin in Co. Mayo, at Collooney, south of Sligo, and on Benmore Head, Co. Antrim, at an elevation of 630 feet above the sea, from which this noble cliff rises precipitously.[2] Those which I have named are referable to the period of general glaciation; but amongst the mountainous districts where local glacial streams originated, such as those of the Reeks and Connemara, they are too numerous to specify.

Travelled blocks, or Boulders.—Another means of determining the direction of former ice-motion is by observing the nature of the large erratic blocks of rock which are often to be found strewn over the surface of the country, or imbedded in the Boulder Clay, or Till, and ascertaining the position of their parent masses. Such blocks are occasionally far too large—several tons in weight—and have been transported over such uneven ground, sometimes up the sides of hills and ridges, to allow it to be supposed that they have been carried by water power. We are obliged to fall back in such

[1] See fig. 26.   [2] See fig. 24.

cases on the agency of ice, either as sheets, glaciers, or icebergs. Such blocks of rock, generally called 'boulders,' resemble, in their position and mode of transport (though on a smaller scale), the large erratic blocks of granite which are strewn along the flanks of the Jura Mountains, high up above the Lake of Geneva, and at a distance of fifty miles from their parent masses amongst the Central Alps. Having identified by their petrological characters the sources of these erratic blocks, we generally find that the direction of transport thus indicated coincides with the ice-movement as determined by any observations that may have been made on surface-scorings or grooves, and thus the one set of phenomena corroborates the other.[1]

I have already referred to some cases of travelled blocks, and may here mention a few others before passing on to another part of my subject.

The upper surface of the hilly district formed of Cambrian and Lower Silurian rocks which lies between the Wicklow Mountains and the sea is often covered by boulder clay, but sometimes the bare

---

[1] This is not always the case, as the transport of the boulders may belong to a different period to that of the ice-sheet or glacier. Thus numerous erratic blocks which have been carried by floating ice during a late stage of the Glacial epoch indicate a direction of transport differing from that of the rock-scorings. Mr. Close has made some very judicious observations on this subject in his paper on 'The General Glaciation of Ireland.'

slaty or quartzite rocks reach the surface. Sometimes resting on the naked rock, at other times lying upon the boulder clay, or imbedded in it, are huge blocks of granite from the interior of the mountains to the westward. Streams of such blocks may be traced up the great glens which penetrate the interior, such as Glenmalure and the Vale of Clara, into Glendalough and Glendasan. Some of these blocks are of large size, occasionally as big as a small cabin, and they must have travelled as far as ten or twelve miles from the districts where the granite first commences on the slopes of Lugnaquilla and Lugduff, or other neighbouring heights. Rising from the centre of the Silurian district, is a remarkable prominent ridge of quartzite called Carrick Mountain, which with its rocky, gnarled, back rises 1,252 feet above the sea; and on the side of this ridge opposite the mountains, we find boulders of granite and schist stranded in large numbers, up to very nearly the top; while (as Mr. Close has observed) glacial striæ pointing in a direction a little north of west may be found on the surface of the quartzite at the northern end of the ridge. Now this direction of the glacial striæ points towards the position of the granite mountains from which the boulders have come, and indicates the carriage of the blocks by means of a glacier or sheet of ice stretching from the interior mountains outward over the hills to the east. The western flank of Dunran Hill, which lies to the north of Carrick Hill, is similarly glaciated.[1]

Again, on the summit of the bare ridge of Cronebane, which is formed of Lower Silurian schists and reaches an elevation of 816 feet, overlooking Castle Howard and the beautiful Vale of Ovoca, there lies a huge boulder of grey granite, measuring 14 feet in length, about 10 feet in height, and nearly as much in breadth. From its side the eye ranges across the richly wooded glens of Ovoca and of the Avonbeg, away up into the wilds of Glenmalure and Glendalough, and then to the cloud-capped heights of Lugnaquilla, Clohernagh, and Kippure—the mountain birthplace of the great wanderer by our side; and we picture to ourselves, how different is the scene now from what it was when those dark masses of mountain which bound the horizon to the westward were clothed in perennial snows, and when the forest-clad valleys at our feet were filled with ice of such depth as to rise over this ridge on which we stand, and of such mass as to carry on its surface this granite monolith, known in all the country round as 'the Mottha Stone,' which the an-

---

[1] Close, Journ. Roy. Geol. Soc. Ireland, vol. i. p. 12. I have recently verified Mr. Close's observation of the direction of the ice-striæ on Carrick Hill (1877).

tiquarian regards as a 'Druidical altar,' and the country folk as the plaything, or weapon, of the traditionary Finn Macoul. Its real history is more wonderful than either, for the facts of nature far outstrip the stories of romance.

Thus, all the evidence goes to show that the ice in this district has moved nearly from west to east;—a direction which is quite independent of the forms of hill and dale, proving (as Mr. Close observes) that the ice-sheet must have been of enormous thickness.

One more illustration of the transport of erratics will suffice for the present, and it will be taken from

FIG. 25.

The 'Clough More.' A granite boulder stranded on slate at a height of 957 feet above Rostrevor Bay. Length, 12½ feet, height, 9 feet.

another mountainous region where glacial action of a general character is remarkably prominent; namely, the northern slopes of the Mourne Mountains.

These slopes rise abruptly from the shores of Carlingford Bay behind Rostrevor Harbour, and on the rounded shoulder which projects northward above this place has been stranded a conspicuous boulder of granite from the district near Newry to the northwards. The elevation of the site of the boulder is 960 feet, and it lies on a surface of Silurian Slate, of which that portion of the hill is composed (see fig. 25). Now, from the character of the granite there can be no doubt it has come from the district which lies to the N. or NNW., and it has been carried across the valley of Rostrevor which opens out

Fig. 26.

A perched block on a glaciated rock near Pontoon Bridge, Lough Conn. The rock is granite, and illustrates the form of 'Crag and Tail.'

upon the lough, and up the hill side to the spot where it has been left stranded. This direction agrees very closely with that of numerous observations of glacial striæ obtained by Mr. Close and Mr. Traill,[1] along the shores of the lough, showing tha

[1] See Geological Survey Map, Sheet 71.

the ice-movement in that district has been in a SSE. direction.

The shores of Lough Conn, in Co. Mayo, exhibit some excellent examples of glaciated rock surfaces in granite, and of travelled boulders and perched blocks, resting sometimes in critical positions, as in the case represented in fig. 26, near Pontoon Bridge.

CHAPTER II.

PERIOD OF GENERAL GLACIATION.

In speaking of the period of the 'General Glaciation of Ireland' it must be understood that I refer to that epoch of intensest cold, represented by the Lower Boulder Clay, the oldest of the Post-Pliocene deposits, to which I have already referred (p. 78). I have shown how widely this formation is distributed over the surface of the older and more solid rocks, and that it is altogether different in its composition and structure from ordinary deposits which have been formed under water.[1] That it is the production of an ice-sheet there can be no doubt whatever. The stones and boulders which are found caught up and imbedded in it, are often quite different in character from those rocks on which it happens to rest, and show that the ice-sheet, when moving across the ancient floor of the land, has broken off and carried away the fragments of this floor, while rubbing and

---

[1] For an account of the formation of the Boulder Clay, see Geikie's Great Ice Age,' 2nd edit. p. 59.

grinding them down either to a smaller size, or even converting them into hardened mud.

Now we have to enquire, what are the laws which governed the movement of the ice-sheet, and what are the directions of this movement? The replies to these questions are to be obtained (as we have seen) by a careful study of the directions of the ice-marks on the rocks over the face of the country, the forms of ice-worn rocks themselves, and (with certain exceptions) the dispersion of erratic blocks. Do such observations, it may be asked, lead us to arrive at any *general* views regarding the movements of this great sheet of ice? This question has received a full and satisfactory answer from the pen of one of the most distinguished of Irish geologists, to whom I have already referred, the Rev. Maxwell Close, in his masterly papers on the Glaciation of Ireland.[1] He has shown *that there exists a tract of country, stretching across the island, which has been the axis of motion for the ice in opposite directions seawards.* This general result, one of the most remarkable in several respects that I know of in the glacial history of the British Isles, has been arrived at by noting and comparing a large number of observations of glaciated surfaces all over the country. There is one feature at least regarding this glacial movement which differs from

[1] See Ref. to Authors, p. 258.

that of any district in Britain, or indeed of Europe, with which we are acquainted. Almost universally the snow-fields which are, or have been, centres of glacial dispersion are situated amongst mountainous tracts, down which the ice has moved on to the valleys and plains. In Scotland, as has been clearly shown by Mr. James Geikie, the ice from the Grampians has moved westwards out to sea, or southwards over the Central Valley, till it met another stream from the Southern Uplands. In England, the mountains of Cumberland and Westmoreland, and in Wales, those of the Snowdon group, have been centres of dispersion. In Europe the great elevated Scandinavian table-land has sent its ice-flood over the low-lying plains of Northern Germany, and invaded the south-eastern portions of England; and farther south, the Alps and Pyrenees have been vastly greater reservoirs of snow and ice than at present, having formerly spread their ice-floods over the adjoining plains and valleys. But in Ireland it was otherwise; for the average elevation of the central reservoir and source of the stream is probably not more than 400 or 500 feet above the sea; and from this the ice moved northwards, and southwards, and westwards, crossing the plains, marvellously regardless of local irregularities of the surface, and only diverted from its rectilinear course by elevated mountain slopes, round which it moved, or along the shoulders of which it sometimes ascended, passing over the lesser elevations of opposing barriers. Thus in a stately, unbroken stream it moved along till it entered the lap of the ocean, in which its wanderings had their natural and peaceful end. Let the reader inspect the little map (p. 224), and he will see that this view of the movement of the ice-sheet is borne out by the observations there recorded.

Cause of the ice-movement.—Such conditions of ice-movement, it will be evident, require some special explanation other than that applied to the motion of ordinary glaciers and ice-sheets. It is not the descent of a plastic mass along the sloping floor of a valley, set in motion by the force of gravity, or impelled by masses of snow pressing downwards from the higher elevations, that we have to deal with; but the movement of a mass of ice from a comparatively depressed region over a plain nearly equally depressed, and, for great distances, over undulating or hilly ground often more elevated than the plain itself. It was a propelling force, owing to which the ice was forced to climb over opposing ridges, the shoulders of mountains, and barriers of rock considerably more elevated than the ground from which the stream had originated.

The position of this central snow-field and source of movement is clearly indicated on the map of general glaciation (p. 224). It occupies the tract lying between Lough Corrib and Lough Mask on the west, and Lough Neagh on the east—a tract coinciding with the northern portion of the Central Plain of Ireland, and bordered along the north by hills and uplands, over which the ice has moved in its efforts seawards. As Mr. Close truly observes, this region owes its existence as a source of the ice-movement to geographical position rather than to relative elevation; and, considering the question from this point of view, I think we cannot be at a loss for an explanation of the phenomena here referred to. Mr. Close suggests, with some hesitation, as a cause of the motion of the ice, that the western side of the country has been somewhat relatively tilted, so as to give the land a higher elevation over the tract now occupied by the counties of Mayo and Roscommon than that lying to the south and east. But the source of the ice-movement has not been from a *focus* in this region; but (as may be fairly deduced from the directions of the ice-movement in Cavan, Monaghan, Armagh, and Antrim) from a line of country, or axis, stretching for a distance of upwards of 100 miles from the WSW. to the ENE. The tilting of the land, therefore, towards the western extremity of the axis would scarcely account for the outward motion of the ice along the centre and east. On the other hand, let us enquire whether we may not discover a cause for the ice-movement of another, and more probable, kind. Is it not easily conceivable that the line of country here described was at the earliest glacial period the region of greatest snowfall—the region of greatest precipitation of snow, as it is now that of greatest precipitation of rain ?[1] Of course, this region included all the tract lying to the northwards and westwards; but the proximity to the ocean (and consequent facility afforded for the dissolution and breaking up of the ice in these directions) limited the area of accumulation *as such* to a comparatively narrow band of country. Nor is it difficult to account for the vast accumulation of snow over the region here indicated. The north-western districts of Ireland directly confront the Atlantic, and are swept by the prevalent westerly winds, which come in from the ocean charged with moisture. The ranges of hills which are bounded by the mountains of Galway in one direction, and by those of Donegal in the other, bring down immense quantities of rain over the region drained by the

[1] Of this, however, I have no positive proof, as the observations on the rainfall in Ireland are not sufficiently general and prolonged to render the matter certain. But from what I know of this tract I have little doubt that it is the case.

head waters of the Suck, the Shannon, the Erne, and the Blackwater, while the mountains on either hand condense these vapours, which return in copious streams to the ocean. Thus, as is probable, the region aforesaid is that of greatest rainfall at the present day, and was that of greatest snowfall during the Glacial period, outside the immediate confines of the several mountainous districts situated along the western seaboard, but which (owing to their proximity to the sea) were drained of their snows too rapidly to become centres of ice dispersion to any great extent. Over this band of country, extending from Galway to Antrim, the snows, we may suppose, were piled up to an enormous depth, and to a less extent over the tracts lying to the southward and eastward, until the *vertical* pressure of thousands of feet of snow and ice was ultimately converted into a *lateral* pressure, forcing the ice to take a direction towards those points where it would find an outlet; namely, the sea-coast bounding the Central Plain, and the old river valleys opening out on the coast between the mountains.

That the vertical pressure of the enormous pile of ice and snow heaped up along the line of country which I have described as the axis of motion gave the initial movement to the ice-sheet, both towards the north and the south, I have little doubt. But when we find that motion propagated to upwards of one hundred miles from its source, and that the force has caused the ice to move not only over plains, but up the sides of opposing hills and ridges, we are obliged to have recourse to other modes of explanation for so remarkable a circumstance. Various theories have been proposed in order to explain the motion of glacier ice, but the greater number only deal with this motion when acting along a sloping plain, or the bottom of a glacier valley, where the force of gravity comes into play. But the phenomena of ice-movement in Ireland is evidently independent of this motive force;—in fact the motion has frequently been in direct opposition to it. On this ground, James Forbes's theory of viscosity, and Dr. Tyndall's theory of fracture and regelation must be considered insufficient, and we have to fall back upon some explanation which involves the idea of molecular force within the ice-mass itself. The idea is shadowed forth in Charpentier's dilatation theory, according to which a glacier is impelled by the force of water, percolating through the fissures in the ice and from time to time freezing, and thus by its expansion forcing the glacier forwards. This view has been more fully developed by Dr. Croll in his able work, 'Climate and Time.'[1] This author observes : 'Ice

[1] Chapter xxi. p. 514. Objection has been taken to Dr. Croll's

is evidently not absolutely a solid body. It is composed of crystalline particles which though in contact with each other, are not packed together so as to occupy the least possible space, but are united to one another at special points determined by their polarity ... It will be obvious, then, that when a crystalline particle melts it will not descend in the manner already described, but capillary attraction will cause it to flow into the interstices between the adjoining particles. The moment that it parts with the heat received, it will of course solidify, but it will not solidify so as to fit the cavity which it occupied when in a fluid state. For the liquid particle in solidifying assumes the crystalline form, and there will be a definite proportion between the length, breadth, and thickness of the crystal; consequently it will always happen that the interstice in which it solidifies will be too narrow to contain it. The result will be, that the fluid particle in passing into the crystalline form will press the two adjoining particles aside in order to make sufficient room for itself between them. The crystal will not form to suit the cavity; the cavity must be made to contain the crystal. And what holds true of one particle

---

views on the ground that he makes use of the term 'molecule,' as a molecule cannot either melt or crystallize. If, however, the term 'particle' be used instead, the objection disappears.

holds true of every particle which melts and solidifies. This process is, therefore, going on incessantly in every part of the glacier, and in proportion to the amount of heat which the glacier is receiving. This internal pressure resulting from the solidifying of the fluid particles in the interstices of the ice, acts on the mass as an expansive force, tending to cause the glacier to widen out laterally in all directions,'[1] and I may add, to move with a linear motion in the direction of least resistance.

As this was a period of general elevation, the coast was somewhat further removed from the central portions of the country than at present; the land gained considerably where the sea was shallow, the ice-sheet protruded into the seas around, while at the same time Ireland was united to Britain. Still, the sea would then lie in the same direction as at present, though more remote from the axis of dispersion; and as the ice moved southwards it would (as in England) gradually melt away where it did not actually enter the sea or the ocean.

Local centres of ice-movement during the period of general glaciation.—While the general movement of the ice over the Central Plain and outwards was governed by the dominating force exerted over the region of the greatest snow accumulation, the several mountain

[1] Ibid., p. 523, 524.

groups and ranges lying to the north, west, south, and east of this region were doubtless themselves centres of local accumulation of snow, and centres of glacial dispersion. Amongst these mountains the vestiges of local systems of glaciers, in the form of masses of moraine matter, are often apparent; but the more recent glaciers of the period of the Upper Boulder Clay, and the overspread of the sea waters during the period of greatest depression which intervened, have more or less obliterated the traces of original glaciation referable to the period of the Lower Boulder Clay. Nevertheless, we may feel sure, *a priori*, that while the great ice-sheet was sweeping over the plains, and impinging upon the flanks of the mountains, such as those of Mourne, Wicklow, and Slieve Bloom, the ice from these elevations descended towards the plains, and there met with, and chafed against, the dominant ice stream; often producing such scenes of chaotic confusion, of piling up of ice masses, and of fracture and crushing amongst the contending sheets, equalling, perhaps, in grandeur those which navigators have pictured and described as occurring during 'a crush' of the ice amidst the Arctic Seas. But it is clear from the marvellous regularity and persistency of the glacial striæ, even when encountering such narrows as that between Slieve Bloom and Maugher

Slieve, or between the Commeragh Mountains in Waterford and Slievenaman, that these local streams were unable to do more than slightly deflect the great stream from its onward course, due to its position with reference to the coast outlet on the one hand and of the source of supply on the other. From all this it will be evident that the depth of the ice must have been very great; meanwhile, it is necessary that I should describe somewhat in detail the evidences in proof of the general movement of the ice-sheet in the manner above described.

CHAPTER III.

DETAILS OF MOVEMENT OF THE GREAT ICE-SHEET.

It is evident that the views I have advanced can only hold good if supported by detailed evidence founded on observations of glacial action in the districts surrounding 'the Central Snow-field,' and I shall, therefore, describe these phenomena over these tracts, availing myself of the observations of Mr. Close, and of other observers who have written on the subject. The region to be described naturally divides itself into two portions, that lying to the north, and that lying to the south of the central snow-field, or axis of motion, representing the areas of ice-movement in opposite directions.

1. *District North of the Central Snow-field.*—This tract includes the Donegal and Mayo Highlands, and is indented by the numerous bays and arms of the sea, such as Clew Bay, Killala Bay, Sligo Bay, Donegal Bay, Lough Swilly, and Lough Foyle, while it is drained by the rivers Moy, the Erne, the Foyle and other streams. With the glaciation of this

district we have a good knowledge except at a few points which require to be filled up, but do not affect the general results. Commencing in the district west of Lough Neagh, Messrs. Nolan, Hardman, and Cruise have determined the direction of the ice-flow to be generally northwards, except when the stream has been deflected by opposing mountain ridges. Thus near Cookstown the direction is N. 10 W., and nearly due north, near Moneymore on the western flanks of Slieve Gallion; but west of this, in the district between Pomeroy and Beragh, and on Carrickmore as determined by Mr. Nolan, the direction is deflected towards the north-east, owing apparently to the presence to the northwards of the high ridge of the Sperrin Mountains, which rise to an elevation of 2,240 feet. The northward movement of the ice in this district is indicated by the form of the rocks at Oritor, which crag to north-east.

The direction of the flow along the valley of the River Foyle near Londonderry has been determined by Mr. Close as north-easterly, and further south at Raphoe, I ascertained, during a recent visit, the direction to be N. 10 W., with the crag northwards; showing that the ice moved down into Lough Swilly, and oceanwards, between the high grounds which bound the lough on either side.

On arriving at the district between Lough Swilly

and Galway Bay, including the Donegal Highlands, we at once encounter evidences of opposing ice-currents due to the fact that we have arrived at a region of local dispersion. For our knowledge of the glaciation of this district we are indebted chiefly to Mr. J. F. Campbell and to Mr. Close. From their observations it would appear that the central snow field of the district stretched from Barnesmore on the south into Glen Beagh, enveloping the granite mountain of Slieve Snaght (or the Snow Mountain). From this region the ice-movement radiated, the ice passing westwards into Gweebarra Bay; while another stream descended along the valley of the River Glen, which on facing the bold prominence of Slieve Leag, which rises abruptly from the Atlantic, was split in twain, one branch 'being shunted' into Teelin Bay, the other finding its way into the ocean along the western base of the mountain.

The accumulation of snow over the ridge of Barnesmore must have been great, as the ice appears to have moved away in opposite directions from its flanks, one stream going down the glen by Stranorlar in an easterly direction to join the main flood coming up from the south, and the other entering the head of Donegal Bay in a south-westerly direction, and causing a westerly deflection in the stream from the Lough Erne Valley.

The direction of the ice-flow along the Lough Erne Valley has been determined at several points. On the flanks of the hills overlooking the valley of the Erne east of Swanlinbar, Mr. Wilkinson and myself found a surface of grit, which indicated a movement in a direction N. 25 E., the crag facing north, showing that we had passed to the northern side of the axis of motion at this point. We also observed well-defined striæ on limestone, at a quarry by the roadside near Enniskillen on the west side of the Erne Valley pointing N. 15 W. Mr. Close indicates a westerly movement of the ice about Ballyshannon, and it is clear the flow bent round to the westward very much in accordance with the form of the Erne Valley itself.

The district of high table-lands and deep valleys lying on either side of Lough Allen, and to the northward, borders upon the axis of movement. Mr. Cruise, of the Geological Survey, has observed striations pointing a little north of west at elevations of about 1,100 feet on Slieve-an-Ierin and the Arigna hills, which is in a direction nearly at right angles to those of the Lough Erne Valley. Here we have apparently two sets of striæ indicating a double motion of the ice. Recently, in conversation with Professor Ramsay, the explanation of this phenomenon dawned upon me. Professor Ramsay had noticed a similar double set of striations amongst the Yorkshire hills; one set, guided by the line of the valleys; the other, independent of these, and occurring on the Carboniferous table-lands above. He supposes in this case that the whole of these hills being enveloped in ice, there was a double simultaneous movement;—one along the valleys, and the other over the upper surface of the country. Applying this explanation to the representative district around L. Allen, we can understand how the ice, endeavouring to move seawards, would have a simultaneous flow along the valley of the Erne, and westwards over the surface of the table-lands of Millstone Grit of Leitrim and Sligo.

The direction of the ice-flow along the tract ranging from Sligo Bay to Lough Conn and the flanks of Nephin has been determined by numerous observations made by Mr. Close, Mr. R. G. Symes, and others, as well from rock-striations as by the transport of erratics. At Collooney, Pontoon, and the eastern shore of Killala Bay, the rocks have been moulded and glaciated evidently by powerful erosion coming from the south-east; while blocks of granite and schist from the Ox Mountains have been strewn over the Carboniferous plain to the northwards. The glaciated bosses of granite and perched blocks along the shores of Lough Conn cannot fail to have struck the observer as remarkably fresh and prominent.

The westward movement of the ice along the shores of Clew Bay, and outwards into the Atlantic along the western coasts of Mayo, have been determined by the officers of the Geological Survey. At 'Old Head,' on the southern shore of Clew Bay, the Silurian rock-surfaces are well glaciated by ice which has moved from E. to W., and are often covered by enormous masses of boulder clay, containing (amongst fragments of other rocks) great blocks of serpentine, torn from the northern slopes of the Croagh Patrick Ridge.[1] On Clare Island, Mr. Symes and myself observed some beautifully delicate glacial lines engraved on a quartz reef, indicating two directions of motion—one a little north, the other a little south, of west. The ice-sheet has evidently protruded outwards over this island, as also over Inishturk, and Inishbofin, whether at that time connected by land to the neighbouring coast, or separated, it is impossible to say. The general ice-flood of the interior received accessions from the precipitation of snow on the mountains both north and south of Killary Harbour, down which a mass of ice of great thickness must have moved westwards out to sea.

[1] The direction of the ice-flow was determined by Professor Ramsay and the Author, on August 27, 1877, at Old Head, near Louisburgh.

The traces of glaciation of the district north of Galway Bay, and extending along the shores and islands of Loughs Corrib and Mask, have been worked out by Mr. Kinahan, who observes that the ice-striæ of Lough Corrib appear to belong to, at least, two distinct systems, one being much older than the other. The older may be called 'the primary,' and seems to have been produced by ice coming from the N.E., that is, from the Central Plain of Ireland; the newer seems to be due to systems of glaciers which had their source in the hills W. of Galway and S.W. of Mayo.'[1] This region was indeed intensely glaciated, both by later local glaciers, and at the period of the early general ice-movement. The ice-sheet coming across from the Central Plain fell in with that descending from the quartzite mountains of Connemara, which caused it to deflect southwards along the northern coast of the bay to such an extent that it extended all over the Aran Islands, where Mr. Kinahan has observed striæ pointing about N. 25 E.[2]

We have now extended our observations not only along the northern, but the western, margin of the great central snow-field, from which it will have

[1] Explanatory Memoir to sheet 85 of the Geol. Survey Maps.
[2] The grinding of the granite apparent along the road to Clifden from the west, referred to by Mr. Close, is probably due to the local glaciation of a later period.

been seen that there was a general outward movement all along this tract. We now have to examine the evidence along, and south of, its southern border, which is not less convincing.

2. *District South of the Central Snow-field.*—That there was a great movement of the ice out of Galway Bay has long been the opinion of Professor King,[1] Mr. Ormsby, and others. This movement extended over the Bay and the Aran Islands, and also over the district south of the Bay. Mr. H. M. Ormsby describes the glaciated surface of the limestone between Galway and Oranmore. The striations were found to be parallel to the direction of the railway, or about E. 20 N., and along with these the rock was deeply grooved in some places.[2] We have here got into the great ice-stream, which was continued in a south-westerly direction along the southern shores of Galway Bay, where Mr. Close records striæ pointing WSW. The observations made at Loughrea, Gort, and Ennis, together with the drift-carriage, point to

[1] Quoted by Mr. Close, to whose paper I must refer the reader for many details of the glaciation of this region for which space cannot be found here. See also Expl. Memoir on Sheets 93, 94, 95, 103, 104 and 105 of the Maps of the Geological Survey of Ireland.

[2] Journ. Roy. Geol. Soc. Ireland, vol. i. p. 19. Mr. Close is convinced that the striæ at Athenry station prove that the ice moved at that place from the WNW. (*supra cit.*, p. 223); but, if so, this must have been due to the extension of the later ice-sheet referred to by Mr. Kinahan, which originated in the Connemara highlands.

a general south-westerly movement of the ice, changing along the northern shore of the Shannon mouth into a due westerly course.

The details of the ice-movement over the tract of country bordering the Shannon Valley will be better understood by reference to the map than by any description. The general result has been worked out by Mr. Close. The ice, tending to move southwards, here met with obstructions to its course by the uprising of the Slieve Boughta, and Slieve Bernagh Hills on the one side, and the Maugher Slieve and Slieve Bloom Mountains on the other. On approaching Slieve Boughta, the most advanced of these barriers, the stream was forced to divide; one branch trended in the direction of the Shannon mouth, and the other was forced to move right across the Shannon Valley near Portumna, and N. of Nenagh, where observations show a SE. course. In taking this direction, however, the stream had to cross the neck which connects Slieve Bloom and Maugher Slieve, over which it passed and escaped into the open plain of Tipperary and Kilkenny.

Part of the stream must have been banked up against the northern slopes of the Commeragh Mountains, or have met the local flow of ice descending these slopes. But the main stream forced a passage through the hollow between these mountains and

Slievenaman, rising over their shoulders and making towards the coast, where it rejoined the stream that had moved southwards along the eastern flanks of Slieve Bloom; and thus it moved onwards towards the coast of Waterford and Wexford.[1] The southerly transportation of detritus observed by Du Noyer in Co. Waterford bears out these conclusions.

Returning now to the southern borders of the Central Snow-field, and east of the Shannon, the striæ and the direction of the ice-drift in the counties of Longford and South Leitrim have been determined by the late Mr. Foot, of the Geological Survey, showing that the movement was towards the SSW.[2] At this spot we are close to the ice-parting, but at the spots indicated on the map the ice had fairly started on its journey southwards. The directions of the drift-ridges east of the Slieve Bloom Mountains flow into and are parallel with the lines of glacial erosion at Carlow, and we now find ourselves in that powerful branch stream which swept the western flanks of the Dublin and Wicklow Mountains[3] as it moved

---

[1] The striations in the neighbourhood of Kilkenny have been observed by Mr. Hardman, of the Geological Survey. They all point a little E. of S. The scorings observed by Dr. Westropp near Kiltorcan and S. of Kilkenny range NNW. and SSE., and those observed by Mr. Du Noyer on the east of the Commeragh Mountains range from the NNW. to SSE.

[2] Journ. Roy. Geol. Ireland, vol. i. p. 32.

[3] At the Chair of Kildare the direction is SSW. (W. H. Baily and A. McHenry.)

towards the south-eastern shores of the island. But to this we shall return.

We now approach the eastern shores, and here some very interesting details of the ice-movement have been worked out by Mr. Close. We have seen how the ice-flood coming from the northward divides into two streams in front of Slieve Boughta. This course is repeated where the ice approached the northern flanks of the Dublin and Wicklow range. The ice coming from the Cavan district, where the striæ indicate a south-easterly course, appears to have divided in the neighbourhood of Maynooth; one branch being deflected nearly at right angles to its normal course, sweeping round the base of the Dublin mountains, and then turning southwards, formed a confluent with the great stream which, as we have seen, swept along the western slopes of the Wicklow range. The eastern branch took a course of a rather complicated character, partly in a SE. direction out into the Irish Sea as it now exists, and partly southwards, in which direction it has been forced over the minor ridges and spurs which form the advanced outworks of the mountains. To anyone who is familiar with the appearance of ice-moulded rocks it will be evident that the granite slopes of Dalkey and Killiney have received their contours from the action of ice moving towards the south-east; that is to say, from ice ascending from the plain and moving over the ridge, the general direction being, according to Mr. Close, N. 43 W.[1] The granite bosses at the corner of Killiney Park, near a small quarry, where they are worn into 'crag and tail,' show similar phenomena. The stream has also ascended and flowed over the quartzite ridge of Shankill, in a direction nearly south, and at an elevation of 912 feet,[2] it has passed down the gorge of 'the Scalp,' and has swept along the flanks of the Greater and Lesser Sugar Loaf Hills, where glacial markings are recorded at elevations of 800 to 900 feet. The stream was here deflected slightly by the uprising of the former beautiful cone. The bold bluff of Bray Head retains on its hard quartzite flanks and summit indubitable evidence that it has been completely enveloped in the ice-sheet. The polished and striated surfaces of the quartz rock may be traced to the very summit of the hill at an elevation of 793 feet, and on several of the minor elevations, the general direction being (according to Mr. Close) S. 31 E., where the flow appears to have been less liable to deflection from surface irregularities.

[1] Journ. Roy. Geol. Soc., vol. i. p. 5. Rock striations in this district have been noticed and recorded by Dr. Oldham.

[2] Close, *Ibid.*

[3] A beautifully ice-worn boss of quartzite may be seen on the summit of the ridge overlooking the town of Bray, covered with parallel lines and grooves pointing S. 20 E., the ice having moved from the north.

On the northern and southern slopes of the ridge, but especially on the latter, the boulder clay has been piled up to a great depth, while the ridge itself was swept bare of covering. Erratic blocks of limestone derived from the north are stranded on the slopes of the hill; so that we have concurrent testimony of several kinds to the southward and eastward direction of the flow. Considering the abrupt character of the ridge, though less so to some extent at that period along the seaward face than at present, the spectacle which the ice-sheet must have presented (had a spectator other than angelic been present to behold), as it forced its way over the great barrier of solid quartzite, schist, and granite that here bounds the Central Plain of Ireland, must have been truly wonderful. The glittering, nearly boundless ice-sheet would (we may suppose) have been seen spread over the plain to the northwards, bearing on its surface freights of boulders and shingle. On approaching the base of the ridge it becomes more and more broken, fissured, and *crevassed* until, on reaching the ascent, the surface is broken up into walls of ice, piled up tier above tier, sometimes regular, at other times heaped up in wild confusion; and thus it surmounts the ridge, and passing over it descends the declivities on the opposite side in a broken surging, but silent, torrent, until reaching the plain and the borders of the coast, it closes up its ranks and passes onwards in its stately path towards the waters of the sea.

All along the eastern shore from Dublin to Dundalk Bay, as well as at Howth, Ireland's Eye, and Lambay Island, the direction of the striæ indicates an eastward movement. North of Balbriggan there is a slight and unaccountable deflection northwards noticed by Mr. Close. The hills north of Dundalk are often intensely ice-moulded, and the 'crag and tail' forms indicate a movement towards the southeast.[1] The narrow valley and lake of Camlough, which lies at the eastern base of Slieve Gullion, coincides in direction with that of the ice-movement. This lough to all appearance is due to glacial erosion, and the ice appears to have swept the drift clean out of the valley, and piled it up at its southern extremity, where it opens out on the plain, and where the depth and pressure must have been lessened and relaxed.

That remarkable groove which is occupied by the Newry Canal and stretches down Carlingford Lough, very nearly coincides with the direction of the ice-flow. Beautifully preserved groovings occur on the limestone surface at Greenore, the direction being about SSE. Carlingford Lough itself is a kind of

[1] These are well shown along the flanks of Fork Hill.

fiord, a long and deep arm of the sea, shallower near the entrance than in the interior. According to the Admiralty charts, the centre of the lough opposite Killowen Point has a depth of 16 fathoms, while opposite the lighthouse at the entrance the maximum depth is 4 fathoms. It is clear, therefore, that were the coast elevated by 4 fathoms, the Lough would be converted into an inland rock-basin with a depth of 12 fathoms near the centre, the outfall being over beds of limestone which crop out on either side of the entrance. Carlingford Lough, therefore, affords a good illustration of those sea loughs, or fiords, so prevalent along the coasts of Scotland and Norway, where glacial erosion has acted so powerfully, and must (as Professor Ramsay has demonstrated) be considered as the agent in wearing out such hollows amongst the harder rocks. Mr. James Geikie has well described these fiords as they occur in Scotland.[1] The northern slopes of the Mourne Mountains frequently show glaciated faces of granite, and Mr. W. A. Traill has recorded several observations of striæ. Here, however, the solid flanks of Slieve Donard have opposed an obstacle to the southerly motion of the ice, and, in consequence, the stream has been deflected, and (as shown by observations along the north flanks of this mountain)

[1] 'Great Ice Age,' 2nd edit., p. 279.

has been obliged to move eastward out to sea. In so doing, however, it has left numerous monuments of its former presence on the flanks of the ridge above Tollymore Park, where the sides of the mountain are strewn with granite boulders up to an elevation of about 1,200 feet above the sea, while the granite rock is ice-worn up to a height of 1,400 or 1,500 feet.[1] The boulder clay ascends to a high level on the slopes of Slieve Donard and its neighbouring heights. The valley of the Bloody Bridge River, which enters the sea south of Newcastle, has been filled in by boulder clay, out of which the existing river channel has been excavated. Ascending by this valley, we find the stony clay, with large erratics, assuming the character of moraine matter, and stretching up the mountain side to the base of the steep slope (about 1,500 feet) of Slieve Donard, which rises 2,796 feet above the sea. Mr. Traill and myself determined the upper limit of the boulder clay on Slieve Muck, which rises 2,198 feet (as determined by the Ordnance Surveyors and exactly corroborated by our aneroid observations!) to be about 1,570 feet.[2] So that we may assume that only the highest

[1] One of these measured 12 feet by 9, and was partially imbedded.

[2] Slieve Donard itself presents a very rounded outline when seen from the north; but it does not follow that this is due to ice-grinding, as granite from its uniform structure has a tendency to weather into dome-shaped masses.

elevations of the Mourne Mountains were left uncovered by the ice-sheet, if indeed there were any such.

It now remains to consider the direction of the ice-flow at the north-eastern extremity of the island, and what was its relation (if any) to that of the opposite coast of Scotland. Over the hilly district of County Down, where the boulder clay forms large rounded ridges, the general direction is east of south. Near the southern end of Strangford Lough at Portaferry, Mr. Traill has recorded striæ pointing SSE. The directions of ice-striæ at or near Tullynakill, Comber, and Castle Espie, recorded by Mr. Close, all have a general SSE. course. The upper surface of the isolated basaltic plateau of Scrabo Hill near Newtown Ards, already referred to,[1] is remarkably ice-moulded, but, as is usual with exposed surfaces of basalt, actual striations are seldom preserved. Nevertheless, they have been observed by Mr. Campbell, Professor Ramsay, and myself, pointing in a direction NE. by N. At Bangor on the northern coast the direction is N. 10 W. There can be no question but that the ice-flow was from the north in this district during this stage. The drift carriage and the bearings of the ice striæ to the north of Slieve Donard all concur in leading to this conclusion.

[1] See p. 51.

Crossing Belfast Lough we reach the basaltic escarpment of Antrim. On the summit of Divis Mr. Hardman observed faint lines on the basalt pointing north and south; a similar direction has been observed on the basalt of Island Magee, corresponding with that of the striations on the Silurian rocks at Bangor, recorded above.

The problem we are now considering derives additional interest from the proximity of this region to that of Scotland, where the ice was accumulated to a depth of several thousand feet, and protruded out to sea from the adjoining coasts. We have seen that to the south of Belfast Lough and of the valley of the Lagan, the general movement was southwards, or a little east of south. This is the view of Mr. Close,[1] Mr. Campbell,[2] and of the officers of the Survey; and corresponds with what we should expect, having regard to the position of the central axis of movement which passed through Lough Neagh.

Over the north-eastern portion of Co. Antrim, the ground is broken by deep valleys opening out upon the sea-coast, and by hills of considerable elevation, so that the uniform outward movement of the ice has been locally interrupted and diverted. In the

---

[1] *Supra cit.*, p. 215. Mr. Close quotes the observations of Mr. Doyle.
[2] 'Frost and Fire,' vol. ii. p. 61.

neighbourhood of Coleraine, Mr. Close records striations in a WNW. direction on basalt; also near Cullybackey he found striations ranging NNW., and he states that these '*rather* look as if the grinding movement was towards the NNW.'[1] This I believe to be a correct interpretation, though on the little map, Mr. Close makes the ice-movement point in an opposite direction. The matter, however, is set at rest by the very clear ice-striations on the basalt at Port Ballintrae, first noticed by Mr. Traill, and lately visited by myself in company with Professor Ramsay. Here, where the boulder clay has been recently removed, the basalt is strikingly glaciated, the groovings pointing NNW., and the surface of the rock showing clearly that the movement has been in that direction. We may assume, then, that along the tract of country stretching to the eastward of the Lower Bann, the general movement has been a little west of north. Further east, however, the deflection of the ice-striations warns us that we are coming within the influence of other than Irish glaciation. The high table-land of Fair Head presents us with phenomena which can scarcely be explained on the beautifully simple theory of a movement in opposite directions from a central axis. Nowhere, perhaps, in Ireland out of the immediate

---

[1] *Supra cit.*, p. 215.

districts of former local glaciers, are the evidences of ice-action so fresh and obtrusive as on the surface of this high basaltic plateau. Everywhere the rocks are ice-worn,[1] and frequently retain the original striations. Perched blocks are abundant, and have often been left stranded in critical positions on the backs of the *roches moutonnées*; while two little loughs, excellent examples of rock-basins, attest the grinding power of the former ice-sheet.[2] There are two sets of striæ, determined by Professor Ramsay, Mr. Traill, and myself; the dominant one ranging about W. 35 S.; the other W. 20 N. The former of these points in the direction of the Mull of Kintyre, the other towards the Wigtown coast. The ice-movement in either case has been westwards, and were it not that the erratic blocks which are strewn over the surface are all of the local dolerite, it might have been suggested that the Scottish ice-sheet had stretched over to the Antrim coast. While the evidence does not appear to favour so extreme an hypothesis, it indicates, I think, the influence which the Scottish ice had upon the ice-movement in this part of the country. It is easy to

[1] See fig. 24.
[2] One of these little rock-basins is called L. Cranagh, from the remarkable cranogue, or site of an old lake-dwelling, which occurs near its centre. This cranogue is enclosed by a wall of well-fitted stones nearly entire, and covered by lichens.

conceive that the ice-streams of both countries must have met in the North Channel;[1] and this being so, it follows that the greater mass of the Scottish ice, estimated by Mr. Geikie at 3,000 to 3,500 feet,[2] would inevitably cause the lesser mass to be deflected from its normal direction. Thus we may suppose that the ice-sheet, moving in a NNW. direction from the central axis which appears to have stretched across Lough Neagh towards the coast about Glenarm, when approaching the coast came within the influence of the mass which filled the North Channel, and so was deflected into a direction at right angles to its course; in other words, towards the west.

We have now followed the course of the ice-flow to the north-eastern coast of Antrim, and have seen how the ice seems to have moved in opposite directions from a central axis. This axis probably terminated not far from the present Antrim coast; but it is interesting to observe how closely it corresponded with an axis of movement which ranged across the Southern Uplands of Scotland from the district of Carrick in Ayrshire, into Selkirk, as shown by Mr. James Geikie. The ice-sheet descending from this axis into the valley of the Clyde met the more powerful glacier coming from the north. Hence the two

[1] See map in J. Geikie's 'Great Ice Age.'
[2] *Ibid.*, 2nd edit., p. 65.

opposing streams were deflected to the east and south-west, and entering the North Channel and the Firth of Clyde, encountered the ice-sheet of Antrim.

We must now return to the South of Ireland, and endeavour to ascertain what became of the great ice-flood as it moved southwards into Limerick, Kerry, Cork, and Waterford.

From the observations of Messrs. Kinahan, O'Kelly, and Wynne, recorded in the 'Explanations' to the Geological Survey maps of the district south of the Shannon below Limerick, there is sufficient evidence, according to Mr. Close, for concluding 'that the great ice-stream in Clare which flowed towards the south-west, sent off a branch at about ten miles south-east of Ennis, across the Shannon, into Limerick and the north of Co. Cork, by Pallaskenry, Ballingarry, and Charleville.'[1]

We may suppose that in this region, the stream from the north, rapidly melting away under the less arctic conditions of the climate, moved towards the base of the Killarney Mountains, and here encountered large masses of ice descending from the Reeks. On approaching the base of these mountains from the west and north, we cannot but be struck with the numerous erratic blocks which strew the surface, where they have not been removed by the

[1] *Supra cit.*, p. 225.

hand of man, indicating an outward movement of the transporting agent. Here, then, at the base of the Reeks, we must take leave of the ice-sheet which we have traced throughout a distance of over 100 miles, from the margin of the Central Snow-field, near the head waters of the Shannon.[1]

The districts of Cork and Waterford afford abundant evidence that the great ice-sheet has extended even to the southern coast of Ireland. The rocks along the shores and islands of Cork Harbour are often overlaid by boulder clay, with angular or rounded blocks of grit and slate, the surfaces of which are often glaciated; a large deposit of such rests on the limestone of Haulbowline Island. The late Mr. Du Noyer records the general direction of the glacial markings along the tract extending from Dunmanway eastward, to be NNW. and SSE.,[2] and we have already seen that there is evidence of a movement of the ice southwards near Charleville and Kilmallock on the borders of Cork and Limerick. It would, therefore, appear that the ice-flood which crossed the Shannon below Limerick, or passed

---

[1] The evidences of local glaciation on a grand scale amidst the valleys of the Reeks are referable probably to a later period represented by the Upper Boulder Clay and the close of the Glacial epoch.

[2] 'Explanation' to Sheet 193 of the Geological Survey Maps, p. 18.

through the gorge now occupied by the river above that city, swept down southwards by the western slopes of the Galty Mountains, and then continuing towards the coast at Kinsale and Cork Harbour it was deflected eastward by the slopes of the Kerry Mountains, from which it received an accession of ice, and finally entered the ocean in a SSE. direction.[1]

[1] Messrs. Jukes and Du Noyer mention the fact that over all the neighbourhood of Kenmare and the south coast of Cork there are traces of a general glacial movement from the NNW. to the SSE., besides the marks of local glaciers in the promontory separating Kenmare River and Bantry Bay. Mr. Close, who quotes the above opinion of these observers, bears testimony to the statement. *Supra cit.*, p. 226.

CHAPTER IV.

DEPTH OF THE ICE-SHEET.

The depth or thickness of the ice-sheet must have varied greatly according to locality and the physical features of the district over which it moved: where its progress was slow, and where it was in any way banked up, the accumulation of ice must have been enormous; for we must not forget, that although the centre of pressure, where the snow accumulated to such a depth as to give motion to all the surrounding mass, lay along a definite tract of country, yet the whole surface was constantly receiving fresh accessions of snow from the atmosphere, which would more than make up for the loss by surface evaporation, and melting along the bottom, which was probably slow and exceptional.

We have seen that there is evidence that on the northern flanks of the Mourne Mountains the ice was piled up to a height of at least 1,500 or 1,600 feet, and was probably higher. In the opposite side of the country, observations have been made by

Messrs. J. F. Campbell, Close, and Kinahan, all tending to show that in the district of West Galway and Mayo the ice overtopped some of the highest elevations. Thus Mr. Campbell has observed that the summit of Shannaunnafeolagh, 2,012 feet, which forms the south-east of the Maum Mountains in Connemara, is glaciated, and marked by striæ running SW. by W., and NE. by E., while Mr. Kinahan states that glaciated rock still survives on the summit of Ben Gowar, one of the Twelve Bins, which is 2,184 feet above the sea level. Mr. Du Noyer has observed the ice-worn character of the rocks amongst the Reeks to an elevation of 2,500 and nearly 2,800 feet, and they may be observed to be glaciated to a greater altitude than ' the Windy Gap' on the coast of the ridge between Kenmare and Killarney, 2,875 feet. But we must not forget that these glaciated surfaces may be referable to the later period of local glaciation, which should always be kept distinct when we are considering questions relating to the mountainous districts of our island. The case of the glaciation of the summits of the mountains above Sneem, north of Kenmare River, at an elevation of 2,200 feet, recorded by Mr. Wynne, appears to be referable to other than local causes. When we come, however, to glacial markings crossing one of the isolated mountain ridges which rise along the margin of the Central Plain, we have evidence regarding the height over which the ice-flood rose of a perfectly reliable nature; and such is afforded by the observations of Mr. Wynne, who found glacial scorings crossing the summit of the Devil's Bit near Templemore at an elevation of 1,583 feet above the sea.[1] With regard to the height to which the stream rose on the flanks of the Dublin and Wicklow Mountains, we may well believe with Mr. Close that it was *at least* 1,000 feet in depth,[2] and probably much more. The question whether it rose and crossed the Wicklow Mountains and flowed out to seaward to the east is still uncertain. It is a fact, however, that enormous boulders of granite from the interior of the mountains are strewn over the undulating country to the eastward, and on the sides and flanks of the ridges. That the ice-sheet passed over the ridge of the Three Rock Mountain, and Lesser Sugar Loaf, Mr. Close has himself shown, and the prolongation of this current would carry it across Carrick Mountain which is glaciated nearly to its summit, 1,252 feet. We may therefore feel confident that if the stream has not crossed the high passes of the Wicklow Mountains to the west, it has come down over the lower shoulders and spurs which form their northern slopes, and has thus passed out to sea.

[1] 'Explanation,' Sheet 135 of the Geol. Survey Maps, p. 26.
[2] *Supra cit.*, p. 231.

CHAPTER V.

LOCAL GLACIAL CENTRES OF A LATER PERIOD.

THAT several of the mountain districts of Ireland have been the centres of local glaciers towards the close of the Glacial epoch I have already stated when describing Moraines.[1] Such evidences of local glaciation have been noticed by Campbell amidst the Donegal Highlands; by King, Close, and Kinahan amongst the mountains of Galway and Mayo; by Tyndal, Jukes, Close and the officers of the Survey amongst the Reeks of Kerry; by Du Noyer amongst the Commeragh Mountains of Waterford; and by Jukes, Close, and others amongst the mountains of Wicklow; and of these I have already described examples in the case of the moraines of Glenmalure, Glendalough, and Glendasan. The glacial phenomena of the north-eastern Highlands of Mourne and Carlingford I consider—notwithstanding the opinion of Mr. Campbell to the contrary—to be due

[1] See p. 101.

not to local glaciers, but to the movement of the general ice-sheet along their flanks and over their shoulders, saddles, and lesser elevations.

To describe in detail the various glacial systems of these local centres would be useless were it possible; but impossible it is, for the phenomena as they occur in each district would require to be, not only described, but mapped out in detail, and this has not yet been done in any published documents. There is here work for geologists yet to do, and unquestionably it is work of a kind to afford the highest interest to the glacialist and the student of nature. Amongst the grand scenery of these mountains he will find all those evidences of local ice-action and dispersion which other British and Continental mountain districts supply. He will find lateral and terminal moraines, where the glacier in its progress or retreat along the valley has paused awhile, and piled up the débris it has brought down from the heights above. He will find blocks which have been carried from a distance left perched in perilous positions on crags, or on the backs of ice-worn bosses. He will find mountain lochs and tarns, which as rock-basins have been scooped out by the eroding action of the glacier descending from the heights above, or which as reservoirs have been dammed up by masses of moraine matter. In other cases he will see the former bed of a lake which has been laid dry where the mountain stream has cut a channel for itself through the old moraine embankment, and so drawn off the waters; and lastly he will be able to compare the effects of these local and restricted glaciers with those of the earlier, and vastly greater ice-sheets, belonging to the epoch of General Glaciation. To the young geologist, who has time as well as strength at command, I cannot conceive any undertaking more calculated to benefit both mind and body, while he will thus be able to add his contribution to the general store of glacial literature. The fact that such investigations are carried on amongst the grandest scenes of nature, gives a zest that is wanting to the study of many other branches of natural history. I can speak from personal experience of the inspiriting nature of this work, having a good many years ago mapped out in detail the glacial phenomena of the southern portion of the Lake District in England; and, if time permitted, would be only too willing to undertake the work I here recommend to others.

As I have already hinted, simple written descriptions of glacial phenomena are not sufficient. The details should be accurately recorded on good maps with hill-shading. The Ordnance maps on the six-inch scale are the best for field work, and from these the details can be transferred to other maps on a smaller scale. In such a manner only can we have a restoration of the glacial features of our country, and of a character commensurate with their interest and importance.

CHAPTER VI.

A CLOSING CHAPTER.

WITH the account of the glacial phenomena of Ireland, and that of its eskers, raised beaches, and river terraces, the work I have undertaken properly ends; but a few words connecting the pre-human with the present age may not be an inappropriate conclusion of our subject. The aboriginal Celtic race has left the records of its presence in many well-known forms, in its cromlechs, cave dwellings, lake dwellings, or cranogues, raths and barrows. We have seen that the rude flint implements of war, or of the chase, are preserved in the gravels which fringe the northern coasts, formed before those coasts were raised as high out of the sea as they are at present; but before his introduction on the scene, species of animals either entirely extinct, or which have migrated to other climes, roved over the plains, or haunted the forests of Ireland.

First amongst these was the mammoth (*Elephas*

primigenius), the remains of which have been found at Dungarvan and a few other places.[1] This great elephant ranged over Great Britain, Central and Northern Europe and Asia, and appears to have migrated, though, perhaps, in smaller numbers, into Ireland. He was accompanied by another huge pachyderm, the hippopotamus,[2] also probably a rare visitor, but one that would find a congenial habitation amongst the lakes and sluggish rivers of the interior. In ruminants the country was probably rich. Two species of the ancient ox (*Bos frontosus*, and the *Bos longifrons*) occur, whose descendants still survive in Chartly Park, Chillingham Park in England, and the forest of Hamilton in Scotland. But it was in the deer tribe that the Post-Tertiary fauna of Ireland was specially rich; for remains of three species of deer have been discovered amongst the caves, peat-mosses, and alluvial deposits of the country.

First, the Red Deer (*Cervus elaphus*), which survives amongst the mountains of Killarney,[3] and till within half a century ago wandered amongst the glens

---

[1] Prof. Harkness, Geological Magazine, vol. vii. p. 253; and Prof. A. Leith Adams, Trans. Roy. Irish Academy, 1876.

[2] Mr. R. H. Scott, 'Catalogue of the Mammalian Fossils hitherto discovered in Ireland,' Geol. Mag., vol. vii. p. 413. Dr. Leith Adams considers the evidence of its existence in Ireland as doubtful.

[3] The Red Deer of Ireland is probably a different variety from that of the Scotch Highlands. In a bog at Bohoe, Co. Fermanagh, the remains of five individuals have been discovered.

of Erris in North Mayo, and the wilds of Donegal.[1] Then the Reindeer (*Tarandus rangifer*),[2] which once spread over the plains of Central England and France, but as the climate became warmer, and unsuited to its comfort, migrated northwards to its present arctic habitation.[3] Last and noblest, was the great Irish Deer [4] (*Megaceros Hibernicus*), which, although an inhabitant of Europe and Britain, seems to have made Ireland the home of his choice, and to have flourished and abounded to an extent not elsewhere known. His remains have been found in almost every quarter of the island, but especially at Lough Ghur in Co. Limerick, and on the slopes of the Dublin Mountains, where they have been recently unearthed in great numbers by Mr. Moss of the Royal Dublin Society. These remains generally occur at the bottom of the peat bogs, where they rest on the white shell marl of the old lakes, whose waters once occupied the position

[1] Professor Owen, 'Palæontology,' gives a drawing and description of a fine antler of this species from the bed of the Boyne at Drogheda, now in the Museum of Sir Philip Egerton, Bart., which measures 30 inches in length, and sends off not fewer than 15 branches, or tynes, including the terminal cluster of tynes, which gave to the deer developing them at the period of his full perfection the title of 'crowned hart.'

[2] Dr. Oldham, Journ. Geol. Soc. Dub., vol. iii. p. 252.

[3] It is supposed that in warm or temperate climates the reindeer is attacked by a fly which is a plague to its existence, but from which it is free amidst the snows of Lapland.

[4] Erroneously designated an 'elk,' but the form and position of the antlers forbid this name.

from which they have been displaced by the encroachment of the vegetation.

I attribute the abundance of individuals of the Great Deer to the absence of many of the natural enemies with which he had to contend in Britain and Europe. In these latter countries he was liable to be seized upon by animals of the feline tribe, such as the hyæna and cave lion, which are not known ever to have entered this favoured isle. Against the ambuscade and stealthy spring of such carnivores, his fleetness would be of little avail, and consequently his numbers were probably kept down in the neighbouring countries. In Ireland, however, he was 'Monarch of the Glen,' and his only enemies were the wolf and the bear, from which the fleetness of his limbs, and his powers of swimming, would afford him a reliable means of escape. From these circumstances, and judging by the number of his remains actually discovered from time to time, we may conclude that this noblest of the deer tribe roved over the plains and valleys of Ireland in large herds, sometimes swimming its lakes and rivers when pursued, but generally leading a life free from danger, till man himself at length appeared on the scene; and to him the final extinction of this noble animal is in all probability due.[1]

---

[1] From a statement in the 'Annals of the Four Masters,' there

The remains of the horse have been found associated with those of the mammoth and bear at Dungarvan as well as in other caves; it is, however, questionable if the horse was indigenous. But it is unquestionable that three species of bear were amongst the early inhabitants of Ireland (*Ursus arctos? U. spelœus,* and *U. fossilis,* sive *ferox*). The physical characters of the island were peculiarly favourable to the existence of such animals, but judging by the paucity of their remains, they do not appear to have been numerous. The woods and marshes were inhabited by herds of wild swine (*Sus scrofa*), doubtless the progenitors of the long-snouted, long-legged Irish pig, now in his turn becoming gradually extinct before the importation of the English variety. Some species of sheep and goat also abounded amongst the rocks and mountains, while the wolf ranged in hungry packs over the country, and only became extinct at a late period.

Some of these animals survived into recent times, as we have seen. The red deer, which only now lives in a wild state at Killarney, has been unhappily exterminated out of the mountainous tracts of North Mayo and Donegal, owing to the destruction of

is some reason to believe that the great deer and man were contemporaries in this country—a view which is entitled to belief on general grounds.

the forests which were once extensive in those districts. The wolf has, *not* unhappily, been exterminated, according to Dr. Scouler, within the last century; while the martin, the wolf-dog, and the wild boar have only disappeared since the end of the twelfth century, at which date we have entered on the period of authentic history.

APPENDIX I.

AUTHORS REFERRED TO.

Baily, W. H., 'Explanations' to Maps of the Geol. Survey. *Passim.*
Bryce, Dr. James, 'On Animal Remains in the Antrim Caves,' Trans. Brit. Assoc., 1834, p. 658.

Campbell, J, F., 'On the Glaciation of Ireland,' Quart. Journ. Geol. Soc., vol. xxix. p. 198.
Campbell, J. F., 'Frost and Fire' (London), 1856. 2 vols.
Carte, A., 'On the Remains of Reindeer which have been found Fossil in Ireland,' Journ. Geol. Soc. Dub., vol. x. p. 103.
Close, Rev. Maxwell, 'On the General Glaciation of the Rocks near Dublin,' Journ. Roy. Geol. Soc. Ireland, vol. i. p. 3.
Close, Rev. Maxwell, 'On the General Glaciation of Ireland,' Ibid., p. 207.
Close, Rev. Maxwell, 'On some Corries and their Rock-Basins in Kerry,' Ibid., vol. ii. p. 236.

Du Noyer, G. V., 'On the Flint-Flakes of Antrim and Down,' Journ. Roy. Geol. Soc. Ireland, vol. ii. p. 169.

Enniskillen, the Earl of, 'Type Specimens of Fossil Fishes,' Geol. Mag. vol. vi. (1869).

Geikie, Professor A., 'Scenery and Geology of Scotland,' 2nd edit.
Geikie, James, 'The Great Ice Age,' 2nd edit. 1877.
Geological Survey of Ireland. 'Explanations' and 'Explanatory Memoirs' to accompany the Maps.
Griffith, Sir R., 'On the Relations of the Sedimentary Rocks of the S. of Ireland,' Journ. Geol. Soc. Dub., vol. viii.
Griffith, Sir R., 'On the Divisions of the Carboniferous Series,' Brit. Assoc. Rep. 1837 (Section C.) p. 88, also Journ. Geol. Soc. Dub., vol. vii. p. 271.

Griffith, Sir R., 'On the Physical Geology of Ireland.' Second Report of the Railway Commissioners (1838).

Hardman, E. T., 'On some New Localities for Upper Boulder Clay,' Journ. Roy. Geol. Soc. Dub., vol. iv. (New Ser.)

Hardman, E. T., 'On the Age and Mode of Formation of Lough Neagh,' Ibid., vol. iv. p. 170.

Harkness, R., 'On the Rocks of the Highlands of Scotland and N.W. of Ireland,' Quart. Journ. Geol. Soc., vol. xv.

Harte, William, 'On the Occurrence of Kjökkenmöddings in Co. Donegal,' Journ. Roy. Geol. Soc., vol. ii.

Haughton, Rev. S., 'On the Fossil Red Deer of Ireland,' Journ. Geol. Soc. Dub., vol. x. p. 12.

Haughton, Rev. S., 'On the Granites of Ireland,' Brit. Assoc. Rep. 1863; Quart. Journ. Geol. Soc. vol. xii. p. 177, and vol. xiv. p. 300.

Hull, Edward, 'Presidential Address to the Geol. Soc. Ireland,' Journ., vol. iv. pt. 2 (1874-5).

Hull, Edward, 'On the Geological Age of the Ballycastle Coalfield,' Journ. Geol. Soc. Ireland, vol. ii.

Hull, Edward, 'On the Structure of Haulbowline Island,' &c., Ibid., vol. iv. p. 111.

Hull, Edward, 'On the Distribution of British Carboniferous Strata of British Isles, as illustrated by Geo-diametric Lines.' Quart. Journ. Geol. Soc., vol. xviii. p. 127.

Jukes, J. Beete, 'On the Mode of formation of some River-valleys in the South of Ireland,' Quart. Journ. Geol. Soc., vol. xviii. p. 378.

Jukes, J. Beete, 'Manual of Geology,' 3rd. edit. (1872).

Jukes, J. Beete, 'Explanations' of the Maps of the Geological Survey of Ireland. *Passim.*

Kane, Sir R., 'Industrial Resources of Ireland,' 2nd edit.

Kinahan, G. H., 'On the Eskers of the Central Plain of Ireland,' Journ. Geol. Soc. Dub., vol. x. p. 109.

Kinahan, G. H., 'On the Drift of Ireland,' Journ. Roy. Geol. Soc. Ireland, vol. i. p. 191.

Kinahan, G. H., 'Supplementary Notes on the Drift of Ireland,' Ibid., vol. iii. p. 9.

Kinahan, G. H., 'Valleys and their Relation to Fissures and Fractures in Faults' (1875).

Appendix I.

Kinahan, G. H., 'On the Estuary of the River Slaney, Co. Wexford,' Journ. Roy. Geol. Soc. Ireland, vol. iv. p. 60.

Lyell, Sir Charles, 'Antiquity of Man,' 4th edit.
Lyell, Sir Charles, 'Principles of Geology,' 3 vols. 3rd edit.

Macalister, Alex., 'Notes on Irish Crania,' Journ. Roy. Geol. Soc. Ireland, vol. ii. p. 188.
Murchison, Sir R., 'Siluria,' 4th edit. 1867.

O'Kelly, J., 'Explanations' of the Maps of the Geological Survey.
Oldham, Prof. T., 'On the more recent Geological Deposits in Ireland,' Journ. Geol. Soc. Dub., vol. iii.

Portlock, General, 'Geology of Londonderry and Tyrone' (1843)

Ramsay, A. C., 'Physical Geography and Geology of Great Britain 3rd edit. (London.)

Scott, Robert S., 'On the Granitic Rocks of S.W. Donegal,' Journ. Geol. Soc. Dub., vol. ix. p. 28.
Scouler, Dr. R., 'On the Lignites and Silicified Woods of Lough Neagh,' Journ. Geol. Soc. Dub., vol. i.

Traill, W. A., 'On Geological Sections in the Co. Down. Rep. Brit. Assoc. 1874, p. 93-95.

Westropp, W. H., 'Sketch of the Physical Geology of North Clare,' Journ. Roy. Geol. Soc. Ireland, vol. iii. p. 75.
Wynne, A. B., 'On some Physical Features of Land formed by Denudation,' Journ. Roy. Geol. Soc. Ireland, vol. i. p. 256.

APPENDIX II.

LIST OF THE CHARACTERISTIC FOSSILS OF THE GEOLOGICAL FORMATIONS OF IRELAND.

CAMBRIAN.

Hydrozoa or *Zoophyta.*—Oldhamia antiqua, O. radiata.
Annelida.—Histioderma Hibernicum, Arenicolites, tracks and burrows.

LOWER SILURIAN.

Hydrozoa.—Didymograptus Murchisoni, Diplograptus pristis, Graptolithus Sedgwicki, G. Hisingeri.
Actinozoa.—Favosites fibrosus, F. alveolaris, Halysites catenularius, Heliolites interstinctus, Cyathophyllum (Petraia) elongatum.
Echinodermata.—Glyptocrinus, Palæaster obtusus, Echinosphærites aurantium.
Crustacea.—Agnostus trinodus, Æglina mirabilis, Ogygia Portlocki, Phacops Brongniarti, Asaphus gigas, Harpes Flanagani, Beyrichia complicata, Cybele verrucosa, Illænus Bowmanni, Lichas Hibernicus, L. laxatus, Remopleurides longicostatus.
Brachiopoda.—Crania divaricata, Discina oblongata, Leptæna quinquecostata, L. sericea, Orthis biforata, O. calligramma, O. elegantula, Strophomena expansa, S. deltoidea.
Lamellibranchiata.—Ctenodonta obliqua, C. radiata, Pleurorhynchus dipterus, P. pristis, Modiolopsis expansa, M. Brycei, Mytilus cinctus.
Pteropoda.—Ecculiomphalus Bucklandi, Conularia elongata.
Gasteropoda.—Cyclonema rupestris, Holopea concinna, Murchisonia turrita, Patella Saturni, Raphistoma elliptica.
Heteropoda.—Bellerophon bilobatus, &c.
Cephalopoda.—Orthoceras bilineatum, O. elongatocinctum, O. subundulatum, Lituites Hibernicum, Phragmoceras compressum.

UPPER SILURIAN.

Hydrozoa.—Graptolithus priodon.
Actinozoa.—Alveolites Bechea, Cyathophyllum elongatum, Cystiphyllum cylindricum, Favosites cristatus, F. fibrosus, F. Goth-

Appendix II. 277

landicus, F. multipora, Halysites catenularius, Heliolites interstinctus, Omphyma turbinata, Syringopora bifurcata.

Echinodermata.—Periechocrinus moniliformis, Actinocrinus pulcher.

Annelida.—Tentaculites Anglicus, T. tenuis, Trachyderma squamosa, Serpulites longissimus.

Crustacea.—Beyrichia Klædeni, Calymene Blumenbachii, Encrinurus punctatus, Illænus Bowmanni, Leperditia Balthica, Phacops caudatus, Prœtus latifrons.

Polyzoa.—Fenestella Milleri.

Brachiopoda.—Atrypa hemisphærica, A. reticularis, Discina perrugata, Leptæna sericea, Lingula Lewisii, L. Symondsi, Orthis elegantula, O. reversa, Pentamerus oblongus, P. undatus, Rhynchonella borealis, R. Llandoveriana, Spirifera plicatella, Strophomena applanata, S. depressa, S. englypha, S. funiculata.

Lamellibranchiata.—Cardiola interrupta, Modiolopsis complanata, Pterinea orbicularis, P. retroflexa.

Gasteropoda.—Euomphalus alatus, E. funatus, E. lautus, Holopella obsoleta, Murchisonia Lloydii, Pleurotomaria undata, Trochus multitorquatus.

Heteropoda.—Bellerophon dilatatus, B. trilobatus.

Cephalopoda.—Orthoceras angulatum, O. annulatum, O. subgregarium.

UPPER OLD RED SANDSTONE. (Kiltorcan.)

Fish.—Bothriolepis (Dendrodus)? Coccosteus, Pterichthys, Glyptolepis elegans,

Crustacea.—Belinurus Kiltorkensis, Pterygotus (or Eurypterus) Hibernicus, Pt. Anglicus, Proraca ris MacHenrici.

Mollusca.—Anodonta Jukesii.

Plants.—Cyclostigma, Palæopteris (Adiantites) Hibernicus, Sphenopteris Hookeri, S. Humphresianum, Sagenaria Bailyana, Knorria Bailyana (fruit).

CARBONIFEROUS SYSTEM.

(*Lower Carboniferous. Marine.*)

Carboniferous Slate, Coomhola grits, &c.,

Polyzoa.—Fenestella antiqua Polypora laxa.

Echinodermata.—Actinocrinus polydactylus, Cyathocrinus planus Platycrinus, Protaster.

Brachiopoda.—Athyris planosulcata, Lingula mytiloides, L. squamiformis, Orthis Michelini, Productus scabriculus, Spirifera striata, Streptorhynchus crenistria.

Lamellibranchiata.—Avicula Damnoniensis, Aviculopecten arenosus, Cucullæa Griffithii, C. Hardingii, C. trapezium, Curtonotus elegans.

Gasteropoda.—Acroculia vetusta, Loxonema, Natica, Macrocheilus.

Heteropoda.—Bellerophon decussatus, B. striatus, B. subglobatus.

Cephalopoda.—Orthoceras undulatum.

Crustacea.—Leperditia subrecta, Phillipsia pustulata.

CARBONIFEROUS LIMESTONE.

(Lower Carboniferous, Marine.)

Actinozoa.—Chætetes tumidus, Cyathophyllum ceratites, C. flexuosum, Lithodendron affinis, L. junceum, Zaphrentis cylindrica, Z. patula, &c.

Echinodermata.—Actinocrinus, Archæocidaris Urii, Poteriocrinus crassus, Crinoids.

Polyzoa.—Fenestella antiqua, F. crassa, F. membranacea, Glauconome pluma, Polypora polyporata.

Brachiopoda.—Athyris ambigua, A. planosculata, Chonetes Hardrensis, Ch. papilionacea, Orthis resupinata, Productus aculeatus, P. cora, P. costata, P. giganteus, P. scabriculus, &c. . Rhynchonella pleurodon, Spirifer glabra, S. lineata, S. cuspidata, S. pinguis, Streptorhynchus crenistria, Strophomena analoga, Terebratula hastata.

Lamellibranchiata.—Aviculopecten arenosus, A. plicatilis, &c. Axinus. Cardiomorpha elongata, Cucullæa tenuistriata, Modiola Macadami, Posidonomya vetusta, Pleurorhynchus Hibernicus, Sanguinolites plicatus, S. transversus.

Gasteropoda.—Euomphalus acutus, E. catillus, E. pentangulatus, Loxonema tumida, Murchisonia sulcata, Pleurotomaria carinata, P. conica.

Heteropoda.—Bellerophon apertus, B. hiulcus, &c.

Cephalopoda.—Goniatites retrorsus, G. sphæricus, G. striolatus, Nautilus biangulatus, N. dorsalis, N. subsulcatus, Orthoceras cinctum, O. Goldfussianum, O. undulatum.

Crustacea. — Brachymetopus Ouralicus, Griffithides globiceps, Phillipsia Derbiensis, P. pustulata.

Fish.—Cladodus mirabilis, C. striatis, Cochliodus contortus, Ctenacanthus tenuistriatus, Ctenopetalus serratus, Deltoptychius acutus, Helodus didymus, Petalodus lævissimus, Pæcilodus Jonesii, Psammodus porosus, P. rugosus, Streblodus Colei, S. oblongus.

YOREDALE BEDS, MILLSTONE GRIT, AND GANNISTER BEDS.

Middle Carboniferous (Marine).

Plants.—Calamites, Lepidodendron, Ferns, &c.
Echinodermata.—Crinoid stems and plates.
Brachiopoda.—Athyris planosulcata, Chonetes Hardrensis, Lingula squamiformis, Orthis resupinata, Productus semireticulatus, Spirifera bisulcata, S. pinguis.
Conchifera.—Aviculopecten papyraceus. A. variabilis. A. alternata, Posidonomya Becheri, P. membranacea.
Gasteropoda.—Loxonema minutissima, Macrocheilus inflata, and var.
Heteropoda.—Bellerophon Urii.
Cephalopoda.—Goniatites sphæricus, G. crenistria, Nautilus cyclostomus, Orthoceras Steinhaueri, O. minimum.
Crustacea.—Phillipsia pustulata, Leperditia Okeni.
Fish.—Cœlacanthus, &c.

COAL MEASURES.

(Upper Carboniferous, Lacustrine and Estuarine.)

Plants.—Calamites cannæformis, Diploxylon elegans, Lepidodendron aculeatum, L. rimosum, L. Sternbergii, Næggerathia, Sigillaria lævigata (alternans), S. (Stigmaria roots), Sphenophyllum Schlotheimii, Sphenopteris latifolia, &c.
Brachiopoda.—Lingula squamiformis (rare).
Lamellibranchiata.—Anthracosia æquilina, Myacites.
Fish.—Cœlacanthus, Helodus, Palæoniscus, Rhizodus.
Reptiles (Amphibia).—Campylopleuron, Dolichosoma, Keraterpeton, Ophiderpeton, Urocordylus.

PERMIAN.

Foraminifera.—Spirillina pusilla.
Actinozoa.—Favosites Mackrothii.
Polyzoa.—Thamniscus dubius.
Lamellibranchiata.—Axinus truncatus (Schizodus), A. Schlotheimii. Bakevellia antiqua, Mytilus squamosus, Pleurophorus costatus.

Gasteropoda.—Rissoa Altenburgensis, R. Gibsoni, Turbo helicinus, I. Thomsonianus, T. Taylorianus.
Crustacea.—Cypridea subrecta.

TRIAS. (*Bunter.*)

Crustacea.—Estheria Portlocki (minuta).
Fish.—Palæoniscus (Dictyopyge) catopteris. (Co. Tyrone.)

RHÆTIC, OR PENARTH BEDS.

Echinodermata.—Ophiolepis Damesii.
Conchifera.—Anomia irregularis, Avicula contorta, Axinopsis Ewaldi, (Axinus cloacinus), Cardium Rhæticum, Cucullæa Hettangiensis, Mytilus Hillanus, M. minutus, Ostrea irregularis, Pecten Valoniensis, Placunopsis Alpina, Pleuromya crassa?
Gasteropoda.—Solarium Thomsoni.
Fish.—Gyrolepis Alberti, G. tenuistratus.

LOWER LIAS.

Actinozoa.—Montlivaltia Haimei, M. mucronata, M. papillata.
Echinodermata.—Cidaris Edwardsii, Extracrinus Briareus, Pentacrinus, Hemipedina Bechei.
Annelida.—Serpula socialis.
Brachiopoda.—Terebratula numismalis, T. punctata.
Conchifera.—Anatina longissima, Astarte Gueuxii, Avicula novemcotæ, Cardinia ovalis, Cucullæa Hettangiensis, Lima gigantea, L. pectinoides, Mytilus Hillanus, M. minimus, Perna, Ostrea arcuata, O. irregularis, Pecten lunularis, Pholadomya ambigua, Placunopsis Alpina, Unicardium cardioides.
Gasteropoda.—Cerithium constrictum, C. decoratum, Chemnitzia Henrici, Cylindrites ovalis, Dentalium Portlocki, Pleurotomaria similis.
Cephalopoda.—Ammonites angulatus, A. Johnstoni, A. planicostatus, A. planorbis. Nautilus.

CRETACEOUS.

Annelida.—Serpula antiquata, S. avita, S. plexus.
Protozoa.—Achilleum fungiforme, Cephalites alternans, C. compressus, Cliona Cretacea, Coscinopora globularis, Scyphia, Siphonia terebrata, Ventriculites alternans, V. decurrens, V. mammilaris, V. radiatus, V. tesselatus, &c.

Appendix II.

Actinozoa.—Parasmilia centralis.
Polyzoa.—Eschara, Holostoma contingens.
Brachiopoda.—Terebratula carnea, T. semiglobosa, Rhynchonella plicatilis, R. limbata.
Lamellibranchiata.—Exogyra columba, Inoceramus Crispi? Lima, Ostrea semiplana, O. vesicularis, Pecten quinquecostatus, Pholadomya decussata, Spondylus spinosus.
Gasteropoda.—Pleurotomaria perspectiva.
Cephalopoda.—Ammonites Lewesiensis, A. Oldhami, A. Portlocki, Baculites anceps, Belemnitella mucronata, Nautilus.
Echinodermata.—Ananchytes ovatus, Cidaris clavigera, C. vesiculosa, Galerites albo-galerus, G. sub-rotundus, Micraster cor-anguinum.
Fishes.—Gyrodus angustus, Lamna acuminata, Otodus appendiculatus, Oxyrhina Mantelli, Ptychodus latissimus.
Reptiles.—Plesiosaurus.

MIOCENE.

Volcanic ash-beds, Ballypaliday.

Plants.—Pinus, Sequoia, Cupressites, Platanus (?) Fagus (?) Andromeda (?) Quercus (?) Rhamnus (?)
Insects.—Beetles (elytra).

PLIOCENE.

Clays of Old Lough Neagh.

Mollusca.—Bivalve shells like Mytilus?

APPENDIX III.

GEOLOGICAL MAPS OF IRELAND.

(1). *The Maps of the Government Geological Survey.*—These are published in sheets on a scale of one inch to the statute mile. The geological lines &c. being engraved on the plates of the Ordnance Survey and coloured by hand. They are issued to the public at a price ranging from 1*s.* to 3*s.* per sheet. The whole of the southern, central, and north-eastern portions are completed.

Small *Explanatory Memoirs* have been prepared to accompany these maps. Some districts are also illustrated by *Horizontal Sections*, on a natural scale of 6 inches to one mile. They are geologically coloured, and are issued to the public at 5s. per sheet.

(2). *Griffith's Geological Map of Ireland*, published in 1838, is on a scale of one inch to four statute miles. It is hill-shaded, and coloured geologically by hand.

(3). *Jukes's Geological Map of Ireland* (new edition by E. Hull, 1878), is on a scale of one inch to eight statute miles. It is geologically coloured by hand.

All the above maps and documents may be obtained directly from the agents, Messrs. Longman & Co., and Mr. Stanford, London; and Messrs. Hodges, Foster, & Co., Dublin

INDEX.

ACH

ACHILL Island, 18
Adams, Dr. Leith, on extinct animals of Ireland, 268
Adiantites Hibernicus, 28
Age of the Donegal and Galway Highlands, 135; of the Wicklow Highlands, 129; of the Killarney Mountains, 139; of the Mourne and Carlingford Mountains, 145
Agglomerate, volcanic, 146
Allen, Lough, 39; origin of, 189; glacial phenomena, 239
Amygdaloid, 66
Animals, extinct, in Ireland, 267-272
Anodonta Jukesii, 28
Anthracite of Kilkenny, 42
Antrim, Co., conglomerates of, 30; volcanic rocks of, 61; raised coast of, 108; direction of ice-movement in, 253
Aran Island, in Donegal, 194; ice-striæ on, 242
Aran Mowddwy, 13
Arches in strata, 159
Ardglass, section of drift-beds, 80; raised coast of, 111
Arigna coal-fields, 153
Arklow, 127
Armagh, Permian beds of, 46
Arra, Slieve, 161
Arrow, Lough, 198
Aughrim river, 127

BER

Auvergne in France, volcanic district, 63
Avonmore river, 127
Avonbeg river, 127

BAGNALSTOWN, 128
Baily, Mr. W. H., on Kiltorcan fossils, 32
Balbriggan, raised coast at, 113
Ball, Mr. V., on volcanic rocks of India, 59
Ballybrack, drift sections, 85
Ballycastle, Co. Antrim, drift terraces, 88; Sea-stacks of, 110
Ballyshannon, direction of ice-movement, 239
Barnaveve Mountain, 143
Barnesmore, 238
Basalt of Antrim, 64
Bays, 208
Beaches, raised, 107, 108
Bear, remains of, 271
Belfast Lough, 50; trap-dykes at, 72
Bellarena, raised beach at, 108
Belmore Mountain, caves in, 205
Benbaun, 15
Bennabeola Mountains, 16
Bent strata due to glacier-ice, 215
Berger, Dr., on the geology of the N. of Ireland, 64

BER

Bernagh, Slieve, 161
Black Valley in Kerry, 194
Blackwater river, its origin and direction, 176, 178
Bole, beds of, 66
Bos frontosus and *B. longifrons*, remains of, 268
Boulder beds of Permian age, 46
Boulder Clay, Lower, 46; position with regard to Pliocene clays, 75; Lower Boulder Clay, 79; its glacial origin, 80, 82, 224; elevation of, 83; like moraine matter amongst mountains, 84; in Co. Wicklow, 218, 234
— — Upper, at Howth Hill, 86; its distribution, 89; thickness near Carlow, 91; mode of formation, 93, 234; in central districts, 245; eastern shore, 246; upper limits in the Mourne Mountains, 251
Boulder stones, 104, 217, 220, 221, 251, 262
Bray Head, 8; raised terrace at, 112; glaciated rocks of, 215, 247
Bray Lough, formed by moraine, 105, 197
Breccia, Permian, of Armagh, 146
Bryce, the late Dr., on remains of extinct animals, Co. Antrim, 110
Bunter Sandstone, 50
Burren country, terraces of limestone, 98

CADER IDRIS, 13
Calciferous Sandstone series, 31
Calp beds, 32
Cambrian rocks, 6
Camlough, glacial phenomena, 249
Campbell, Mr. J. F., on glacial action in Donegal, 194, 238; on depth of the ice-sheet, 261

CLE

Cape Clear, 136
Carboniferous beds, 30, 73, 151, 154; original extent, 155; composition of, 163; denudation of, 166
— slate, 30
— limestone, 30, 31, 144, 151
Carlingford Lough, 12, 249; dykes near, 71; ice-striæ, 249
— Mountains, 12; geological age, 145
Carlow flags, 38
Carmoney Hill, Co. Antrim, 69
Carntual, 132
Carrickareede Island, cave at, 110
Carrickfergus Castle, 73
Carrick Mountain, Co. Wicklow, glaciated rocks, 215, 219; boulders on, 219-262
Castlecomer coalfield, 39, 42, 150, 153, 162
Castlemartyr, 177
Castlerock, raised beach and sand dunes, 109
Caves, old sea-caves of Co. Antrim, 109
Caverns in limestone, 203, 205
Celtic tribes in Ireland, remains of, 113, 267
Central axis of ice-movement, 226, 228, 256, &c.
Central Plain. *See* Plain.
Centres, local, of ice-movement, 233, 263
Cervus alces, C. elaphus, remains of, 268
Chair of Kildare, 12, 150; ice-striæ, 245
Chalk formation, 53; chalk-flints, 55; chalk-cliffs, 56; original extent of beds, 56; origin, 58
Chert in Carboniferous limestone, 33
Clare Island, 18; glacial striæ, 241
Clays, Pliocene, 72, 187; thickness of, 74
Clew Bay, 15, 18; ice-striæ, 241

CLO

Close, Rev. M., on drift-ridges, 81; on glaciation, 82, 95, 211; on drift-gravels, 86; on erratic blocks, 101; on general ice-movement, 225, 228, 242, 252
Coal-measures, 40, 41, 151
Coleraine, ice-striæ near, 254
Colin Glen, Rhætic beds of, 52, 53
Cong, Co. Galway, fountains at, 202
Conglomerates of Upper Silurian age, 23, 125
— of Old Red Sandstone age, 27, 29
— of Pliocene age, 74
Conn, L., 15, 198; glaciated rocks, 223
Connemara, the Twelve Bins of, 15, 18, 261; terraces in, 107
Contraction of the earth's crust, 132
Coomanassig, Lough, 195
Coomhola grits, 30
Cores of Silurian rocks, 160, 172
Cork Harbour, Boulder clay in, 83, 181
Co. Cork, ice-striæ in, 258
Corrib, Lough, a rock-basin, 196, 199, 206
'Crag and Tail,' 216
'Cranogues,' 267
Cretaceous beds, 52
Croagh Patrick, 17
Croll, Dr. J., on the cause of glacier motion, 231
Cronebane, Boulder stone on, 220
Croob, Slieve, 12
Crookhaven, 136
Cross-striations on rocks, 102
Cruise, Mr. R. J., on ice-movement, 237, 239
Crumlin, River, section in, 74, 76
Cuilcagh Mountain, 38, 161, 171
Culdaff, raised coast of, 108

DUN

Cultra, Permian beds of, 48
Cumberland, former ice-action in, 226

DELPHI in W. Mayo, terraces at, 97
Denudation of the Central Plain, 44, 48, 164, 174, 202; of Antrim basalt-sheets, 63; of Carboniferous strata, 140; of SW. of Ireland, 159
Depression of land in Drift Period, 89, 94
Depth of the great ice-sheet, 260–262
Depth of the ice-sheet, 235, 260
Derg, Lough, 171; a rock basin, 196, 200
Derry, Mountains of, 10, 19
Devil's Bit Mountain, 10, 137; glacial striæ on, 262
Dingle promontory, Upper Silurian beds of, 24; Dingle beds, 26; Dingle Bay, 208
Discordance between 'Dingle beds' and Old Red Sandstone, 26
Domes, strategraphical, 159, 160, 172
Donabate, limestone of, 12
Donegal, Mountains of, 10, 19, 162
— ice-movement in, 236, 238
— centre of ice-movement, 263
Douce Hill, 126
Drift-deposits, 78; divisions of drift, 79; ditto in Lancashire, 88; at Killarney, 135; over Central Plain, 151
Duff Hill, 126
Dublin Co., 12; estuary of the Liffey, 112; Dublin Bay, 208
Dundalk, raised beach at, 112, 208
Dundrum Bay, raised coast of, 111
Dungannon, Carboniferous beds of, 39
Dunluce Castle, 56, 66, 69

Dunmanus Bay, 136
Du Noyer, Mr. G. V., on the Dingle Beds, 26; on worked flints, 111; on ice-striæ, 245, 258
Dunran Hill, glaciated rock-surface, 220
Dykes, Basaltic, 51, 70, 72, 142, 145

*E*LEPHAS *primigenius*, remains of, 267
Erne, Lough, and Valley, 29; a rock-basin, 196, 198, 200; direction of ice-flow, 239
Erosion, glacial, 192
Erratic blocks, Co. Mayo, 92; in Galway, 100, 217, 240
Erriff Valley, 22
Errigal Mountain, 20, 162
Erris, Barony of, 179
Escarpments, Carboniferous, 32
Esker ridges, 82, 98

*F*AIR Head, 62; ice-worn rocks of, 216, 254
Faults in strata, Co. Cork, 178; under Lough Neagh, 188; at outlet of Lough Allen, 189, 190; at Swanlinbar, 191
Felspathic rocks, 144
Felstone, 142
Fiords, 249
Flagstones, Carboniferous, 38, 42, 154
Flexures in Lower Silurian beds, 12
— in Metamorphic rocks, 20, 124
— in Carboniferous rocks, 31, 43
— in mountains of Kerry, 130–133, 158,
— in SW. and N. of England, 139
Flint implements at Larne, 110; at Kilroot, 111
Florence Court-Park, 203, 204

Foot, Mr. F., on direction of ice-motion, 245
Foraminiferæ, 166
Fossils, characteristic, Appendix.
Foy, Slieve, intrusive rocks, 144; crystalline marble of, 146
Foyle, River and Lough, 20, 21; drift-gravels of, 88; ice-striæ near, 237

*G*ALLION, Slieve, 12
Galtymore Mountain, 10, 15, 137, 259
Galway Bay, 208; ice-striæ, 242
Galway, West, 10, 15; ice-movements, 263
Gannister Beds, 39, 42, 154
Gap of Dunloe, 133
Gap of Barnesmore, 182
Geikie, Prof. A., on the rocks of the Scottish Highlands, 19; on the raised coast of Scotland, 108, 113
Geikie, Mr. James, on ice-movement in Scotland, 226; on Scottish Fiords, 250; on the thickness of the ice-sheet in Scotland, 256
Geological maps, 5
Ghur, Lough, remains of deer at, 269
Giant's Causeway, 61
Gill, Lough, a rock-basin, 196, 198
Glacial action in Permian Period, 48; in later periods, 192, 213
Glacial evidences, 211
Glaciated rock-surfaces, 81, 133, 197, 212–217; cross-striations, 102, 239, 242
Glaciation, general, 212, 224, 225
— local, 212, 233
Glaciers, extinct, in Ireland, 234, 263

GLE

Glen of the Downs, 182
Glenarm, old sea-caves at, 110
Glendalough in Connemara, 105; in Co. Wicklow, 115, 197, 219
Glendasan, glacial action in, 197
Glengarriff, terraces at, 107, 137
Glenmalure, terminal moraines in, 104; river terraces, 115; erratic blocks, 219
Gneiss, 20
Godwin-Austen, Mr. R. on the formation of the Old Red Sandstone, 27
Grainger, Rev. Dr., on shells of raised beach, 111
Grampian Mountains, 157
Granite of Mourne, 12, 141
—— Donegal, 20
—— Galway, 16
—— Wicklow and Wexford, 12, 128, 162
—— Ox Mountains, 18
—— Slieve Croob, 141
Greenore, raised beach at, 111
Greensand, Upper, 53
Griffith, Sir R., 5; on the Irish coal-fields, 40, 41
Gweebarra valley, 20; ice-striæ in, 238
Gypsum in Keuper marls, 50

HARDMAN, Mr. E. T., on Carboniferous chert, 33; on Pliocene Clays, 74-5; on Upper Boulder clay, 91; on the origin of Lough Neagh, 186; on ice-movement, 237, 245, 253
Harkness, Prof., on the rocks of Donegal, 20; on the epoch of metamorphism, 23; on shelly gravels in North of Ireland, 87
Haughton, Rev. Dr., on Mourne granite, 142; on the volcanic district of Carlingford, 145; on altered Carboniferous limestone of Slieve Foy, 146

INT

Highlands of Scotland, 19, 20, 21, 22; north-western highlands of Ireland, 123, 263; south-eastern highlands, 126, 263; south-western highlands, 130; north-eastern highlands, 141
Hillsborough, trachyte near, 65
Hippopotamus, remains of, 268
Howth, Cambrian rocks of, 6, 7; shells in drift-gravel, 85
Hull, Prof., 3; on Carboniferous chert, 33; on Permian beds of Armagh, 48; on volcanic rocks of Antrim, 59; on the age of the flexures of the Carboniferous rocks of England, 140
Hypersthene dolerite, 143

ICE-ACTION on rocks, 212-216
Ice-movement, east of Wicklow Mountains, 221; general, 225; its cause, 227, 230; Forbes's theory, 231; Tyndall's theory, 231; Charpentier's theory, 231; Croll's theory, 231; details of ice-movement, 236; N. of central snow-field, 236; at River Foyle, 237; Lough Erne, 239; Donegal, 238; Lough Allen, 239; Sligo, 240; Clew Bay, 241; Galway, 242; S. of central snow-field, Galway, 243; Shannon valley, 244; Commeragh Mountains, 244; Waterford and Wexford, 245; eastern districts, 246; eastern coasts, 249; in north-eastern coast, 252; in south of Ireland, 257-9
Inagh, Lough, 16; Glen Inagh, 105
Inishbofin, ice-striæ on, 241
Inishowen, raised coast of, 108
Inishturk, ice-striæ on, 241
Interglacial beds, 84; elevation on Wicklow Mountains, 86

288   Index.

INV

Inversion of strata at Killarney, 135
Iron, Lough, 199
Iron ore from Co. Antrim, 66
Irvingstown, geological section near, 29
Island Magee, old Sea-stacks of, 109, 110

JONES, Prof. T. R., on foraminifera from the Chalk, 54
Joyce's country, 124
Jukes, the late Prof., on the inversion of strata at Killarney, 136; on the origin of river-channels, 169, 177, 180
Jurassic beds absent in Ireland, 164

KANE, Sir R., on the Irish coal-fields, 40, 41; on the Shannon valley, 175
Keishcorran, 182
Kelly, Mr. John, on drift-gravels, 86
Kenmare Bay, terraces at, 107
Kerry, bays and inlets in, 208
Keuper beds, 50
Kilkeel, river, terraces in, 97
Kilkenny, drift-deposits at, 90; marble-quarries at, 90
Killala Bay, 31, 202, 240
Killaloe, gorge of the Shannon near, 200
Killarney, Mountains of, 31, 161, 217, 257, 261; centre of ice-movement, 263
— Lakes of, 133, 194, 199
Killary Harbour, 15; terraces in, 97, 107; glaciated rocks of, 106, 181, 241
Killenaule coal-field, 40, 41, 150, 153, 162
Killiney Hill, 8; glaciated rocks of, 81; Upper Boulder clay of, 89; ice-striæ on, 247
Kilroot, flint implements at, 111

MAC

Kinahan, Mr. G. H., on the Irish coal-fields, 41; on terraces, 98; on eskers, 99; on ice-striations, 242, 257, 261
King, Prof. W., on Permian beds, 48; on the ice-movement in Galway, 243
Kippure Mountain, 126
Knockkroe in Co. Limerick, 36
Knockmealdown Mountains, 177

LACUSTRINE denudation, 206
Lakes, of Killarney, 133, 194, 199; of mechanical origin, 186, 190; of glacial origin, 186, 192, 194; of chemical solution, 186, 198; of the Central Plain, 198, 209
Landslips, 62
Larne, Rhætic beds, 52; raised beach at, 110
Lasaulx, Prof. A. von, on Tridymite, 64
Lee, River, 177, 178
Liffey, River, 127; estuary of, 112
Lignite in Pliocene clays, 73
Limerick, Carboniferous traps of, 35, 36
Limestone, Carboniferous, 30, 31, 150, 154; Silurian, 21; solubility of limestone, 201, 209
'Limestone gravel' of Drift-Period, 84
Lingula flags of North Wales, 8
Lissoughter Mountain, 16
Llandeilo beds, 9, 11
Llandovery beds, 23, 123
Londonderry, 15
Lough Neagh, 185; position of axis of ice-movement, 256
Lugnaquilla, 126

MCDONNELL, Dr., on remains of extinct animals, Co. Antrim, 110

Index. 289

Macgillicuddy's Reeks, 132
'Marble Arch,' Co. Fermanagh, 203
Marls, freshwater, 209
Mask, Lough, a rock-basin, 196, 199, 201
Maugherslieve, 121, 137, 162
Mayo, Mountains of, 10; ice-movement in, 236
'Meeting of the Waters,' 127
Megaceros Hibernicus, remains of, 269; its abundance in Ireland, 270
Melvin, Lough, 198
Mesozoic formations, 50
Metamorphic rocks, 11, 14, 19, 123
Metamorphism, epoch of, in the W. and NW. of Ireland, 22
Microscopic sections of Carboniferous limestone, 32; Chert, 33; Chalk, 54
Millstone grit, 38, 154
Miocene beds, 59, 187
Miocene trap rocks, 45
Mizen Head, 136
Moraines, lateral,
— terminal, 196, 197
Moraines, 101, 102; their composition, 103; lateral and terminal, 103
Mountains, 117; 'the birthday of,' 119; mountain groups of Ireland, 121, 148; of West Mayo and Galway, 123; of Wicklow, 126; of Kerry, Cork, and Waterford, 130, 137; of Mourne and Carlingford, 141; mountains of North Wales, 157
Mountain limestone, 30, 32
Mourne Mountains, glacial phenomena, 221
Mourne Mountains, 12; trap-dykes of, 71
Muckross Abbey, 133
Muilrea Mountain, 17, 18, 24, 124
Muntervary Head, 136
Murchison, Sir R. I., on the rocks of the Scottish Highlands, 19

NAHANAGAN, Lough, how formed, 197
Neagh, Lough, 61; Pliocene Clays, 72; 'old Lough Neagh,' 76; origin of, 185, 186
Nephin, 18
New Red Sandstone, 50; of England, 139
Nolan, Mr. J., on ice-movement, 237

OBLIQUE bedding in gravel, 85
O'Kelly, Mr. J., on the Irish coal-fields, 41; on erratic blocks of Slieve Bloom, 100; on ice-striations, 257
Old Red Sandstone, 26, 27, 137, 177, 180, 190
Oldham, Prof. T., on shells from drift, 87
Oorid, Lough, glaciated rocks at, 105
Ormsby, Mr. H. M.; on ice-striations, 243
Owenbarra valley, 20
Owenmore river, Co. Leitrim, 171
— — Co. Mayo, 179
Ox Mountains, position of, 15, 18; Drift-deposits of, 92; erratics from, 240

PALÆOLITHIC flint implements, 110
Palæozoic formations, 1
Parallelism of mountain chains, 129, 139
Perched blocks, 222; at Lough Conn, 240
Permian beds, 46, 140, 165
Pitchstone of Co. Antrim, 64
Philips, the late Prof., on the age of the flexures of the N. of England, 140
Plain, Central, of Ireland, 120; origin of, 149, 163; its limits, 149-156, 172; lakes of, 196, 198

U

PLA

Plane of marine denudation, 157, 158, 160, 174
Plant beds of Antrim, 67; of Pliocene age, 75
Pliocene clays, 72, 187
Pontoon, 217
Porphyry of Carlingford, 144
Portlock, General, quoted, 61, 88
Port Ballintrae, direction of ice-movement, 254
Port Fad Mine, Co. Antrim, 68
Portrush, Rhætic beds, 52; raised beach at, 109
Post-glacial deposits, 96
Pyroxenic rocks, 144

QUARTZITES, 15, 16, 17, 20
Queenstown Harbour, 177

RAMSAY, Prof. A. C., on the age of the rocks of the Scottish Highlands, 19; on the origin of river valleys, 169; on the glacial origin of lakes, 193; on a double motion of ice, 239; on rock basins, 250
Raphoe, ice-movement at, 237
Rathdrum, bent strata at, 215
Red deer, 271
Ree, Lough, 171, 199
Reptilian remains of 'the Jarrow coal,' 42
Rhætic beds, 51, 52
Ridges of Boulder clay, 81
Rivers, 168
River terraces, 107
River valleys, old, 182
Roches moutonnées, 216, 255
Rock-basins in Connemara, 105, 192; in Kerry, 194, 195; in Central Plain, 196; Carlingford Lough, 250; in Co. Antrim, 255
Rock-surfaces, glaciated, 215

SUG

SANIDINE felspar, 65
Sands and gravels of Post-Pliocene age, 84
'Scalp, the,' 182; its origin, 183; glacial striæ, 247
Scotland, geological formations of, 10, 11; raised coast of, 107; ice-movement in, 256–257
Scouler, Dr., on shells from drift-gravel, 85
Scrabo Hill, 51; ice-worn rocks, 252
Sea-stacks of Co. Antrim coast, 109, 110
Serpentine of Connemara, 16
Seven Churches, ruins of the, 115
Shane's Hill, Co. Antrim, 69
Shannon, River, original direction, 161; source and channel, 171, 175, 190; rainy district of source, 230
'Shannon Pot,' 172
Sheelin, Lough, 199
Shells from Drift-gravels, 86
Silurian rocks, Lower, 8, 9, 137, 141
—— Upper, 21
Slaney, River, 127
Sleamish, Co. Antrim, 69
Slieve-an-Ierin, 153, 161, 239
Slieve Bingian, 143
— Bloom, 137, 162, 172
— Boughta, 10, 224
— Croob, 141
— Donard, 142; ice-striæ on, 250
— Slieve Foy, 143
— More in Achill, 18
— Partry, 18, 124, 161
— Snaght, 162, 194, 238
Sligo Bay, 32; ice-striæ, 240
— Co., 35
Sneem, Co. Kerry, glacial action at, 195
Snow-fall, region of greatest in glacial period, 230
Somersetshire coal-field, 139
Sperrin Mountains, 21, 237
Sugar Loaf Mountain, 7

Index. 291

SYE

Syenite of Carlingford, 144
Symes, Mr. R. J., on ice-movements, 240, 241

TABLE of formations, 3
Tableland of Castlecomer, 42; of Slieve Partry, 158, 174
Tarandus rangifer, or reindeer, remains of, 269
Tardree Hill, Trachyte of, 65
Tate, Prof., on the Rhætic and Lias, 52
Terraces, 32; of Co. Antrim, 63; in Mountains, 96; river-terraces, 113; along the Boyne, 114; Erriff, 114; Glendalough (Co. Wicklow), 115; Co. Sligo and Fermanagh, 203
Tertiary beds, 59
Three Rock Mountain, boulders on, 262
Topley, Mr., on the origin of river-channels, 169
Trachyte porphyry of Antrim, 64
Traill, Mr. W., on trap-dykes of Mourne, 71; on direction of ice-striæ, 222, 250, 252, 254; on upper limit of Boulder-clay, 251
Trap rocks, Lower Silurian, 13
Triassic beds, 50, 187
Tridymite in trachyte, 65
Troughs in strata, 136
Tyrone coal-field, 41, 153

UNDERGROUND river-channels, 202, 203

YOR

VALLEYS in Co. Wicklow, 127
Vartry, River, 127
Volcanic ash, 61, 64
— craters absent in Co. Antrim, 63
— necks, 56–58
— rocks of Carboniferous age, 35
— — of Miocene age, 59
— — Silurian age, 13, 24

WALES, former ice-action, 227
Waterford, trap-rocks of, 13; ice-striæ in, 258
Wenlock beds, 24
Westropp, Dr., on direction of ice-striæ, 245
Wicklow, raised coast at, 113
Wicklow Mountains, 162; slaty rocks bent over at surface 215; a centre of glacial movement, 263
Wilkinson, Mr. S. B., on ice-movement, 239
Wolf exterminated in Ireland, 272
Wright, Mr. J., on Microzoa from the Chalk, 55
Wynne, Mr. A. B., on glaciation in Kerry, 195; on ice-striations, 257, 261

YAUGHAL Harbour, 177
'Yoredale Beds,' 38; at Lough Allen, 189

LONDON : PRINTED BY
SPOTTISWOODE AND CO., NEW-STREET SQUARE
AND PARLIAMENT STREET

MAPS OF THE BRITISH ISLES

PUBLISHED BY

EDWARD STANFORD,

55 CHARING CROSS, LONDON, S.W.

BRITISH ISLES.

STANFORD'S STEREOGRAPHICAL MAP of the BRITISH ISLES. Constructed to show the Correct Relation of the Physical Features. Size 50 inches by 58; scale, 11½ miles to an inch. Mounted on Roller, varnished, 21s.

The method employed in the construction of this Picture of the British Isles is that known as the Stereographic, or the art of representing solid bodies on a plane. In educating the eye to a correct perception of the superficial features of the land, it is necessary to use a symbol as nearly representing nature as the conditions of art will allow, which is accomplished through this method by *imitating vertical relief*, and producing upon the eye the impression of a model.

STANFORD'S OROGRAPHICAL MAP of the BRITISH ISLES. Edited by Professor RAMSAY, LL.D., F.R.S., Director-General of the Geological Surveys of the United Kingdom. Scale, 11½ miles to an inch; size, 50 inches by 58. Mounted on Rollers and Varnished, 30s.

GEOLOGICAL MAP of the BRITISH ISLES. By Professor RAMSAY, LL.D., F.R.S., Director-General of the Geological Surveys of the United Kingdom. Scale, 11½ miles to an inch; size, 50 inches by 58.

STANFORD'S GEOLOGICAL MAP of the BRITISH ISLES. Compiled under the Superintendence of E. BEST, H.M. Geological Survey. Scale, 25 miles to an inch; size, 23 inches by 29.

WALL MAP of the BRITISH ISLES. Constructed on the basis of the Ordnance Survey, and distinguishing in a clear manner the Cities, County and Assize Towns, Municipal Boroughs, Parliamentary Representation, Towns which are Counties of themselves, Episcopal Sees, Principal Villages, &c., with the names systematically engraved according to population. The Railways are carefully laid down and coloured, the Hills correctly delineated, and the Map from its size is well suited for Public Offices, Institutions, Reading Rooms, Railway Stations, good School Rooms, &c. Scale, 8 miles to an inch; size, 81 inches by 90. Price, Full-coloured, Mounted on Linen, on Mahogany Roller, and Varnished, £3.

ENGLAND AND WALES.

GEOLOGICAL MAP of ENGLAND and WALES. By Professor RAMSAY, LL.D., F.R.S., and G.S., Director-General of the Geological Surveys of Great Britain and Ireland. This Map shows all the Railways, Roads, &c., and when mounted in case, folds into a convenient pocket size, making an excellent Travelling Map. Scale, 12 miles to an inch; size, 36 inches by 42. Fourth Edition, with Corrections and Additions. Price, in Sheet, £1. 5s.; Mounted on Linen, in Case, £1. 10s.; or on Roller, Varnished, £1. 12s.

STANFORD'S OROGRAPHICAL MAP of ENGLAND and WALES. Edited by Professor RAMSAY, LL.D., F.R.S., Director-General of the Geological Surveys of the United Kingdom. Scale, 8 miles to an inch; size, 50 inches by 58. Mounted on Roller, Varnished, 30s.

London: EDWARD STANFORD, 55 Charing Cross, S.W.

MAPS.

ENGLAND AND WALES—*continued.*

STANFORD'S LIBRARY MAP of ENGLAND and WALES.
Scale, 5 miles to an inch; size, 72 inches by 84. Coloured £2. 12s. 6d. Mounted on Linen in Morocco Case, £3. 13s. 6d.; or on Roller, Varnished, £4. 4s.; Spring Roller, £6. 6s.

Projected from the Triangulation made under the direction of the Hon. Board of Ordnance, and comprising all the Railroads and Stations; the principal Roads, Rivers, and Canals; the Parliamentary Divisions of Counties; the Site of nearly every Church, distinguishing the Nature of the Living; the Seats of the Nobility and Gentry; also the Distance in Miles and Furlongs of each City and Town from the General Post Office; London.

LARGE-SCALE RAILWAY and STATION MAP of ENG-
LAND and WALES, in 24 Sheets. Constructed on the basis of the Trigonometrical Survey. By ARROWSMITH. On this Map will be found laid down the Rivers, Mail, Post, Bridle, and Railroads, Boundaries of Counties, Gentlemen's Seats, Woods, Covers, &c., as well as the distance from London of all the great Towns. The Railways, with their names, and the names and position of the Stations, are printed in red, thus making them very distinct. The 24 sheets of this Map, being sold separately, will be found extremely convenient and useful for Tourists. Each sheet is 20 inches by 28. Plain, 1s.; Coloured, 1s. 6d.; Mounted in Case, Plain, 2s. 6d.; Mounted in Case, Coloured, 3s. The size of the complete Maps, 114 inches by 128. Plain, in Case or Portfolio, 25s.; Coloured, in Case or Portfolio, 38s.; Mounted on Cloth to fold, in Case, Coloured, £4. 4s.

WALL MAP of ENGLAND and WALES. Scale 8 miles to
an inch; size, 50 inches by 58. Price, Mounted on Mahogany Roller, Varnished, 21s.

This Wall Map is an accurate reduction of the Ordnance Survey, drawn with the utmost precision. The Hills are correctly delineated, the names systematically engraved according to population, and the Episcopal Sees, County and Assize Towns, Municipal Boroughs, Parliamentary Representation, and Towns which are counties of themselves, severally distinguished. The Railways are also carefully indicated and coloured.

SCOTLAND.

STANFORD'S OROGRAPHICAL MAP of SCOTLAND. Edited
by Professor RAMSAY, LL.D., F.R.S. Director-General of the Geological Surveys of the United Kingdom. Scale 8 miles to an inch; size, 34 inches by 42. Mounted on Rollers, Varnished, 18s.

STANFORD'S LIBRARY MAP of SCOTLAND. Scale 5
miles to an inch; size, 52 inches by 76. Coloured, £2. 2s.; Mounted on Linen, in Morocco Case, £3. 3s.; or, on Roller, Varnished, £3. 13s. 6d.; Spring Roller, £5. 5s.

In constructing this Map of the Northern Portion of the British Isles, its mountainous character is well displayed, and whilst showing clearly the minor details, the General Topography of the Country is fully preserved. The Counties are distinctively coloured. The Parishes, Mountains (with their height in yards), the Forests, Hills, Lochs, Rivers, and Canals, the Railways and their Stations, the Mail Roads, the Turnpike and other Roads, the Cities, Market Towns, Villages, Kirks, Manses, Parks, and Ruins are plainly noted, as well as the Islands and Lighthouses around the Coasts. The whole corrected to the present date.

MAP of SCOTLAND, in COUNTIES. With the Roads,
Rivers, &c. By J. ARROWSMITH. Scale, 12 miles to an inch; size, 22 inches by 26. Sheet Coloured, 3s.; Mounted in Case, 5s.

WALL MAP of SCOTLAND, showing the Divisions of the
Counties, the Towns, Villages, Railways, &c. Scale, 8 miles to an inch; size, 34 inches by 42. Price, Full-coloured, Mounted on Linen, on Mahogany Roller, and Varnished, 12s. 6d.

London: EDWARD STANFORD, 55 Charing Cross, S.W.

MAPS.

IRELAND.

STANFORD'S OROGRAPHICAL MAP of IRELAND. Edited by Professor RAMSAY, LL.D., F.R.S., Director-General of the Geological Surveys of the United Kingdom. Scale, 8 miles to an inch; size, 34 inches by 42. Mounted on Roller, Varnished, 18s.

GEOLOGICAL MAP of IRELAND. By JOSEPH BEETE JUKES, M.A., late Director of H.M. Geological Survey of Ireland. This Map is constructed on the basis of the Ordnance Survey, and coloured Geologically. It also shows the Railways, Stations, Roads, Canals, Antiquities, &c., and, when Mounted in Case, forms a good and convenient Travelling Map. Scale, 8 miles to an inch; size, 31 inches by 38. On Two Sheets, 25s.; Mounted on Linen, in Case, 30s.; or on Roller, Varnished, 32s.

STANFORD'S WALL MAP of IRELAND. Scale, 5 miles to an inch; size, 43 inches by 58. Coloured, Four Sheets, 25s.; Mounted in Case, 35s.; on Roller, Varnished, £2. 2s.; on Spring Roller, £4. 4s.
This Map is reduced from the Ordnance Survey, and Coloured in Counties, with the hill-features printed in a brown tint. It shows the Baronies, Towns, Railways, Stations, Roads, Canals, &c. It can also be specially prepared to exhibit the Judicial, Ecclesiastical, Military, or other Divisions of the Country.

MAP of IRELAND. Scale, 4 miles to an inch; size 66 inches by 81; Coloured, £1. 11s. 6d; Mounted on Linen, in Morocco Case, £2. 12s. 6d.; or on Roller, Varnished, £3. 3s.; Spring Roller, £5. 5s.
Originally constructed to accompany the Report of the Railway Commissioners, and subsequently corrected and improved by the addition of important information.

STANFORD'S MAP of IRELAND, in COUNTIES and Baronies, on the basis of the Ordnance Survey and the Census, and adapted to the various branches of Civil and Religious Administration, with the Towns distinguished according to their population, and the Railways, Stations, Roads, Canals, Parks, Antiquities, and other features of interest. Scale, 8 miles to an inch. Size, 31 inches by 38. On Two Sheets, price, Coloured, 8s.; Mounted on Linen, in Case, 10s. 6d.; or on Roller, Varnished, 15s.

MAP of IRELAND, in COUNTIES. With the Roads, Rivers, &c. By J. ARROWSMITH. Scale, 12 miles to an inch; size, 22 inches by 26. Sheet, Coloured, 3s.; Mounted, in Case, 5s.

WALL MAP of IRELAND: showing the Divisions of the Counties, all the Towns, Principal Villages, Railways, &c. Scale, 8 miles to an inch; size, 34 inches by 42. Price, Full-Coloured, Mounted on Linen, on Roller, Varnished, 12s. 6d.

THE PHYSICAL GEOLOGY AND GEOGRAPHY OF GREAT BRITAIN.

By Professor A. C. RAMSAY, LL.D., F.R.S.,
Director-General of the Geological Surveys of the United Kingdom.

Fifth Edition, enlarged. Post 8vo. with Geological Map printed in Colours, and numerous Illustrations. [*Nearly ready.*

London: EDWARD STANFORD, 55 Charing Cross, S.W.

www.ingramcontent.com/pod-product-compliance
Lightning Source LLC
Chambersburg PA
CBHW021954220426
43663CB00007B/807